AF477657

# MAXIMIZING HUMAN INTELLIGENCE DEPLOYMENT IN ASIAN BUSINESS

*Also by Frank-Jürgen Richter*

THE EAST ASIAN DEVELOPMENT MODEL

INTANGIBLES IN COMPETITION AND COOPERATION
  (*edited with Parthasarathi Banerjee*)

ADVANCES IN HUMAN RESOURCE MANAGEMENT IN ASIA
  (*edited with John B. Kidd and Xue Li*)

# Maximizing Human Intelligence Deployment in Asian Business

## The Sixth Generation Project

Edited by

John B. Kidd

Xue Li

and

Frank-Jürgen Richter

palgrave

First published 2001 by
PALGRAVE
Houndmills, Basingstoke, Hampshire RG21 6XS and
175 Fifth Avenue, New York, N.Y. 10010
Companies and representatives throughout the world

PALGRAVE is the new global academic imprint of
St. Martin's Press LLC Scholarly and Reference Division and
Palgrave Publishers Ltd (formerly Macmillan Press Ltd).

ISBN 0–333–94814–9

This book is printed on paper suitable for recycling and made from fully managed and sustained forest sources.

A catalogue record for this book is available from the British Library.

Library of Congress Cataloging-in-Publication Data
Maximizing human intelligence deployment in Asian business: the sixth generation project / edited by John B. Kidd, Xue Li, and Frank-Jürgen Richter.
    p. cm.
  Includes bibliographical references and index.
  ISBN 0–333–94814–9 (cloth)
  1. Personnel management—Asia.  2. Strategic planning—Asia.  I. Kidd, John B.
  II. Li, Xue, 1963–   III. Richter, Frank-Jürgen.

  HF5549.2.A75 M39 2001
  658.3'0095—dc21

                                                              200102737

10  9  8  7  6  5  4  3  2  1
10 09 08 07 06 05 04 03 02 01

Printed and bound in Great Britain by
Antony Rowe Ltd, Chippenham, Wiltshire

# Contents

## PART III   MAKING SENSE OF THE 'US AND THEM' SCENARIO

# List of Figures

# List of Tables

# Notes on the Contributors

**Allan Bird** is Eiichi Shibusawa-Seigo Arai Professor of Japanese Studies at the College of Business Administration, University of Missouri-St Louis. He has worked and lived overseas for eight years in Japan and has travelled and worked in many Asian countries. His research and consulting focus is on international HRM, particularly managerial effectiveness in international organizations and the performance of Japanese multinational companies. He has authored numerous articles and books. His most recent book, co-edited with Schon Beechler, is *Japanese Multinationals Abroad: Individual and Organizational Learning* (Beechler and Bird, 1999).

**John Bratton** is Associate Professor at the University of Calgary, Canada. He received his B.Sc. (Economics) degree from the University of Hull, his MA from the University of Leeds, and his Ph.D. (Industrial Relations) from the University of Manchester, England. He is author of *Japanization of Work: Managerial Strategies for the 1990s* (1992), author and designer of an interactive CD-ROM, *Leadership in Organizations* (1999, AB Access), and co-author of *Human Resource Management: Theory and Practice* (1999, 2nd edn, Macmillan Business).

**William M. Czander** obtained his Ph.D. in Organizational Psychology from New York University. He completed postdoctorate research at the Institute for Social and Policy Studies, Yale University and researched psychoanalysis at their Postdoctoral Center for Mental Health. He is currently Professor of Management in the School of Business at Manhattan College.

**Cherlyn Granrose** (Cherry) received her Ph.D. from Rutgers University and has held tenured faculty positions at Temple University School of Management and Claremont Graduate University School of Organizational and Behavioral Sciences. Her current position is Professor of Management and Organizational Behavior at Berry College, Mt Berry, Georgia. She is an active member of the International Division, and has served on the executive boards of the Gender and Diversity Division and the Careers Division of the Academy of Management. She has been the recipient of Fulbright awards to South Korea, Taiwan, Singapore and the People's Republic of China. Her previous books include *Work–Family Role Choices for Women in Their 20's and 30's*, co-authored with E. A. Kaplan (1996); *Careers of Business Managers* *in East Asia* (1997); and co-edited with S. Oskamp, *Cross Cultural Work Groups* (1997). She has written extensively on Asian human resources management and on women's work and family decisions.

**Mohan Raj Gurubatham** is Consultant and President of Akasha Networks in Malaysia. His previous management positions have included appointments with Pointflex Sdn Bhd., Malaysia; SAP-AG; and Andersen Consulting. He has taught Human Resource Management at the Malaysian Institute of Management, Malaysia; and at the Royal Melbourne Institute of Technology, Australia.

**Harald S. Harung** is President of Harvest AS (a management consultancy) and Harvest Construction AS (a real-estate development firm). He has been a researcher and consultant on the development of human performance for more than twenty years, and has published in numerous journals. He is the author of *Invincible Leadership: Building Peak Performance Organizations by Harnessing the Unlimited Power of Consciousness* (1999, Maharishi University of Management Press), and is an editor of *The Learning Organization – An International Journal*. He has lectured worldwide, and his research on world-class performance has received support from both private and public sources. He has a Ph.D. from the University of Manchester, UK.

**Dennis P. Heaton** is Professor and Chair of the Department of Management and Public Affairs at Maharishi University of Management, Fairfield, Iowa. His teaching activities include delivering MBA courses by distance education at five locations in India, in conjunction with the Maharishi Institutes of Management in Bangalore, Chennai, Hyderabad, Lucknow, and New Delhi. He has a doctorate in Educational Leadership from Boston University, and is the author of numerous chapters and articles on higher stages of individual and organizational development. He is also Assistant Director for the Center for Natural Medicine and Prevention, which is conducting large-scale research projects using traditional approaches from Maharishi Vedic Medicine to enliven inner intelligence for holistic health.

**Ronald L. Jacobs** is Professor of Workforce Development and Education at the Ohio State University, USA. Dr Jacobs also has an appointment as distinguished visiting professor in the School of Business, Nanyang Technological University, Singapore. He is a frequent contributor to the human resource development scholarly literature, having written nearly 100 journal articles and book chapters on human resource development, structured on-the-job training (OJT), employee expertise, and organizational improvement. Dr Jacobs has extensive consulting experience in both manufacturing and service organizations, having worked for numerous large organizations, nationally and internationally. He currently serves as the editor of the *Human Resource Development Quarterly*, the leading scholarly journal of the human resource development field, having over 12,000 subscribers internationally. In 1994, he received the instructional technology research award from the American Society for Training and Development, and in 1995 he was recognized for his scholarly contributions to the field by the Academy of Human Resource Development.

**John B. Kidd** was educated in the UK and worked for several major UK organizations before returning to university scholarship. In the Universities of Birmingham and now Aston Business School his research focused on the development of IT use in SMEs; the management of projects; and the softer management issues that concern multinational joint ventures. At one time he was looking only at Japanese/European ventures, but now focuses on Asian/European ventures. He has held visiting professorships in several European Universities, and in the China Europe International Business School, Shanghai.

**Dong Hwan Lee** is Associate Professor of Marketing and the Gabriel Hauge Fellow in the School of Business at Manhattan College. He received his Ph.D. in Business from Indiana University in Bloomington. Prior to joining Manhattan College, he taught at the State University of New York at Albany and Indiana University in Bloomington. His areas of research include sociological influences on family consumption behaviour, consumer information processing and decision processes, cross-cultural managerial issues, and consumer satisfaction. His research has appeared in *Organizational Behaviour and Human Decision Processes, Psychology & Marketing, Journal of Business Research, Advances in International Marketing, Advances in Consumer Research, Journal of Consumer Satisfaction* and *Dissatisfaction and Complaining Behaviour*, as well as others.

**Xue Li** was educated in Guangxi, China and was in the first group of five persons to be authorized to tutor the Teaching of English as a Foreign Language (TOEFL), and the Business English [Cambridge] Courses (BEC) – indeed she introduced BEC to China. She gained first-hand knowledge of the issues facing managers who jointly wish to set up ventures in China whilst working as a 'go between' in an import/export agency; and also through working as an interpreter for English-speaking firms looking for business in China.

**Philip Merry** delivers workshops and consults on international/intercultural business leadership and team building all over Asia with leading multinationals. He has also worked with a range of well-known educational establishments. He is Visiting Fellow at Bristol University (UK) where he teaches cross-cultural business skills to international MBA students. He has taught at the East West Center – University of Hawaii, National University of Singapore, and is Lecturer on Strategic Alliances at the Melbourne Institute of Technology MBA. He has been an elected member of the governing council of SIETAR – the Society for Intercultural, and Educational Training and Research – which is a global intercultural body affiliated with the UN.

**Joyce S. Osland** is Associate Professor of Organizational Behaviour at the University of Portland. She received her Ph.D. at Case Western Reserve University. She authored *The Adventures of Working Abroad* (1995, Jossey-Bass), as well as two textbooks (with D. Kolb and I. Rubin), now in their seventh edition, *Organisational Behavior and Experimental Approach* (1994, Prentice Hall) and *The Organisational Behavior Reades* (1995). Currently she researches and consults in the areas of global leadership, women leaders, cultural sense making, and Latin American culture. She has won several awards for both teaching and research.

**Frank-Jürgen Richter** is Director for Asia at the World Economic Forum, and has lived, worked and travelled extensively throughout Asia. He has previously held executive management positions with multinational companies, and is a frequent speaker on issues related to Asian economies, international management and global competition.

**Lionel F. Stapley** is the Director of OPUS (Organization for Promoting Understanding of Society) and an organizational consultant. OPUS is an educational charity and company limited by guarantee which is concerned with 'promoting and developing the study of conscious and unconscious organizational and societal dynamics through educational activities; research; consultancy and training; and the publication and dissemination of these activities for the public benefit'. As a consultant, he works with a range of clients at an individual, group and organizational level. Has been a staff member of several international Group Relations Training conferences.

**Oliver C. S. Tzeng** was born in Taiwan and completed his college education in his home country. He received his masters degree in School Psychology at University of Wisconsin, and his Ph.D. in quantitative psychology and cross-cultural social psychology at University of Illinois. Further he studied law and received his J.D. at the Indiana University School of Law – Indianapolis. His research has focused on international modernization, social problems, cultural meanings, inter-personal and inter-cultural relations, comparative laws, and international business and negotiations. He is now Professor of Psychology, Indiana University – Purdue University at Indianapolis, a position he has held since 1976. He is also a practising attorney in Indiana.

# Part I

# Globalization, Strategic Management, and the Deployment of Human Intelligence

# 1 A Perspective on Human Intelligence Deployment in Asia

John B. Kidd, Xue Li and
Frank-Jürgen Richter

## SPACES OF CULTURE, SPACES OF MANAGEMENT

We all have a tendency to mix and work with people similar to ourselves –
because of upbringing, background, education or work experiences. Despite
the omnipresence of the supposed post-modern culture-workers, on the verge
of becoming the prototypes and forerunners of global capitalism, it is still
rather natural to have an 'us and them' mentality. But today, with global
mergers and acquisitions being commonplace, we must open our consciousness to concepts, rituals, and behaviours that are different from our own. As
the world of business is becoming more *globalized*, our everyday applications
are more *localized*: they are fragmented into niches of diversity and interpersonal resistance (Pucik, 1992; Trompenaars, 1993). The pace of change
determines that we will enter into this dichotomy almost directly in our
jobs, and certainly in our daily lives (Kanter, 1995).

Take for instance a mundane task like shopping. The world's goods are
becoming increasingly available in the shopping malls and supermarkets that
are appearing in all major cities, even in the Peoples Republic of China
where once only the family-run shop was found. One may track some of
these changes by noting the M&A (Mergers & Acquisitions) activity of firms
like Carrefour, Metro, or Wal-Mart as they acquire or organize joint ventures in countries far from their origin in Europe or America. As we refocuse
from the family-run business to observe the supermarket, we see they have
to work within outreached global supply chains who have contracts with
growers, manufacturers and logistics groups to put fashionable goods or
fresh products on to the supermarket shelves with speed and precision, so
customers will be enticed to buy goods 'in season' or 'in fashion' anywhere,
not just in Rio, Sydney, Tokyo or London. However, the small grocery shop
in Shanghai or Sulawesi will still have their place in this dynamics of crossing
borders. They will contest and challenge the social stratification of a unified

world, deploying instead trust and personal attachment to customers, suppliers, and communities. The multiplicities of cultural spaces define the background of multinational firms wanting to increase markets shares around the world. As Sony and Coca-Cola, for instance, claim to be *multilocal*, not particularly *multinational*, so the spaces of management have to be carefully re-examined and developed throughout this dichotomy of an evolving world economy.

In this book we are not directly concerned with the details of finance, manufacturing operations or logistics as they are enacted round the world. But we are concerned to open the eyes of managers who must deal with the people involved in these ventures who come from different backgrounds and with whom one must strike a deal, and continue to work in harmonious ways. To do this we need to explore these diversities among people, not as individuals (this we take for granted, even in our own milieu) but in more general terms. We have to consider, for instance, the Western versus Eastern ways of living and working. We once saw 'imperialism' in the form of Western management methods and economic policies being thrust upon apparently willing Eastern firms (Reich, 1991; Gilpin, 2000; Richter, 2001). Yet those local managers were willing, it seemed, 'to put up with whatever it took' in order to be able to import new technology, though they still had to deal with their own local workers. And the local workers have their own natural ways of working, which in the case of China, in particular, could be based upon thousands of years of sophisticated development using rules of behaviour emanating from Confucian teaching. Thus we find interface management of particular importance as the expatriate manager and his local colleagues have to come rapidly to a cross-culture rapport – one to propose and the other to refute arguments – but each in their own way. To be clear, we are not talking only of the Western manager venturing in Asia, but also Asian managers attempting to work across their own national boundaries. Crossing boundaries is enticing for a variety of reasons – market reach can be extended, or new raw materials extracted – but there is the complication always of working with others and coming to a compromise, each with the other, for joint benefit (Richter, 1999).

We may illustrate the centrality of human resources management in the global context by reference to Figure 1.1. Herein we note that the chief executives (CEOs) should be concerned with long-term strategy – where their firm may be heading in the future, trying to out-guess the waves of the economy.

However, in this firm, with its CEO looking maybe 25 years into the future, we see the need for him or her to confer with the more junior staff – the directors, the middle managers and so on, down to the shop-floor workers.

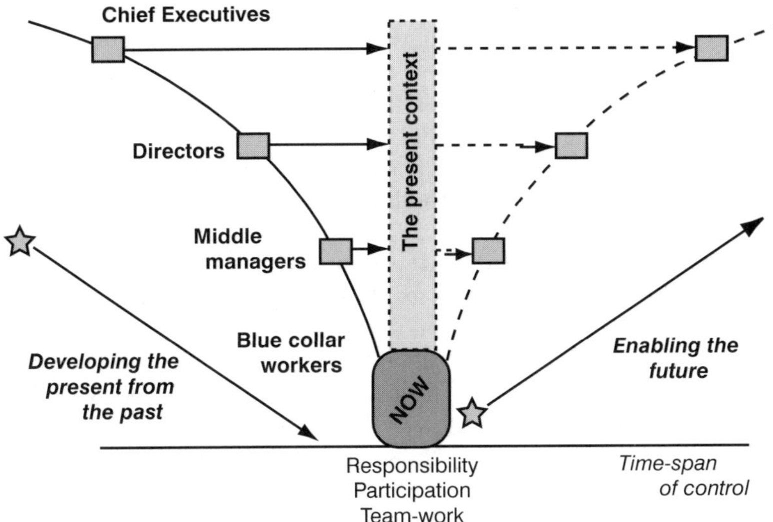


*Figure* 1.1    Views of the future
*Source*:    Following Elliott Jaques (1989) *Requisite Organisation* (London: Casson Hall).

These discussions would, ideally, centre on the firm's strategy, its market place and how new machines, technology and so on would have to be embraced by the workers (at all levels) so as to be competitive. As Figure 1.1 illustrates, it is the shop-floor workers who, in making the goods and in being given the responsibility to develop quality and timeliness of delivery, enable the future of the firm. Meanwhile all management are looking to their respective futures. Yet Figure 1.1 represents an idealistic view – now, with the pace of change being rapid, and as we said above with M&A being so commonplace, all firms must have in place a strong HRM (Human Resource Management) function that will guide their staff development. To do this, the HRM group must also consider the hiring (and firing, in a sense) that will need to be enacted as M&A bring new opportunities to gather staff from joint ventures into one's own firm.

A dynamic HRM function must also seriously consider the knowledge management and organizational learning that has to take place rapidly in the now global firm with its far-flung alliances. How to align the concepts and personal wishes of their multinational staff is a big question, especially as there will inevitably be the 'us and them' attitudes at work.

This book attempts to reach to the managers of multinational firms in their quest for a better way to engage their staff in overseas assignments. Under the present waves of Merger and Acquisition many firms have to achieve several goals (sometimes in conflict), one of which is the great need to manage effectively the ever-leaner and divergent workforce (Kidd and Richter, 2001). Aligning with other firms round the world is fraught with difficulties, but the situation abounds with opportunities – if only we could perceive them. Here we suggest that by knowing better the other, and noting our authors' various perspectives (both East and West), we can help develop this awareness of the others.

Over 2000 years ago Confucius proposed 'five rules of engagement', between ruler–minister, father–son, husband–wife, elder and younger brother, and friend–friend, and so his cardinal relationships and their accompanying moral code have established the importance of hierarchy that survives across Asia to this day. However, there is one relationship that he missed – that between the insider and the outsider (Backman, 1999; Haley and Haley, 2000; Richter, 2001). Outsiders do not fit readily into the Confucian hierarchy; indeed, they represent a challenge to it. Relationships outside this hierarchy are not regulated and are seen as a zero-sum game rather as a win-win game. The Chinese applied war stratagems (*Bing jia*) to monitor their environments that were perceived as rather hostile, and indeed somewhat hostile to Confucianism. The best known strategist was Sun Tzi, a Chinese general who lived in the times of Confucius. His work *The Art of War* (see Sun Tzi, 1996) is a sort of guidebook on how to fight outsiders. Applied to business, Asian managers often treat competition as warfare. If relationships fall outside the five cardinal ones, then war strategies often regulate the interaction between the business parties.

It is this missing sixth relationship that is of most concern to Western business people, because in Asia the expatriate Westerners are one of the most extreme forms of outsider. Effectively, Confucius left an ethical vacuum: how to deal with outsiders and strangers? It is not surprising that Confucius did not address this issue since, in his day, there would be very few true 'outsiders' so he did not establish a further 'rule of behaviour'. Thus, considering the modern Chinese person, who is normally a rule-follower, this person is confused when there is no internalized rule to be followed or broken in confronting an outsider. And with respect to the outsider, he or she maintains a deeply suspicious stance.

Looking more broadly – in the presence of the outsider, Asian people may engage in silent behaviour, such as *Fù Yŭ* in China – keeping words in the stomach before uttering them too soon: 'When a word has left the mouth, not even the swiftest horse can catch it' (Chinese proverb). So, lacking a 6th Confucian rule for 'dealing with the outsider', they prefer silence to making errors and losing face, or conversely in actively promoting the loss of face of the outsider. Similarly, a Japanese person might remember one of their national proverbs – '*Kiji mo nakazuba utaremai*' ('Even a pheasant will not be shot if it keeps quiet') (Takashima, 1985).

We must contrast this Asian behaviour, essentially tacit, with that which developed in the West, also thousands of years ago. The Greek philosophers distinguished between four types of knowledge:

- episteme (abstract, generalizable)
- techne (capability, or the capacity to finish tasks)
- phronesis (practical and social wisdom)
- metis (conjectural intelligence)

From these early concepts the search for 'truth' developed which, in the West through the 1930–1980s, was embodied in laboratory experiments upon one's ability to learn, and the experimenters' skill within the rational scientific method (Détienne and Vernant, 1978; Raphals, 1992). In the later years researchers noted that human judgement was perhaps less bounded, and somewhat irrational, tending to use heuristic rules. Nevertheless, there is still a strong feeling that Aristotelian logic or Euclidean geometry should prevail against intuition and the utilization of tacit skills or knowledge. It follows that Western managers rely on the rule by law, and upon contracts that are enforceable by their laws, while in Asia, with laws that are weaker and irresolute (being enforced by the power brokers – hence their Rule of the Law), managers will rely on honour, on their interpersonal skills, on their *guanxi* (Chinese: relationships; connectedness in human relationships). We are led resolutely, in conducting business East to West, and vice versa, into modes and methods that are inherently different in their roots, but are at the same time enacted between people.

We must become more aware of others in order to better understand the disposition of the HRM function in out-reached firms. In distant out-reaches the local staff have to work closely with expatriates and with buyers and sellers from afar. All have different agendas, so we have to be able to cope with these through the management and deployment of local staff – in other words, to go beyond current HRM practice and to engage in the 6th Generation of Human Intelligence Deployment (HID). In this sense we are noting, through the expressions of our authors, and developing beyond the old established values which are still taught today and which are prevalent across much of Asia, namely the values of Confucianism, Daoism, Chinese Buddhism, Maoism and so on (From and Holmgren, 2000). But we are saying more in this book, since Confucius only developed five strong precepts and ignored how to deal with outsiders: herein we address this issue. Hence the idea of the 6th Generation Project! Here we suggest that the management of knowledge is paramount – but we each treat this aspect differently according to our past learning. To leave this as casual learning is hardly sufficient in the multinational enterprise (MNE) of the present day, given that we see frequent breakdowns in inter-alliance communications. Disunity and discord will not do in the multinational of the future. Even in the near future we expect to see much more use of virtual reality – in the form of quickly made alliances to satisfy niche markets, maybe on the other side of the globe: and once that niche market is satisfied the alliance will be discontinued. There will be virtual teams set up to manage these 'virtual' alliances, and sometimes there will be a need for a 'rapid reaction force' to fly round the world as problem solvers. Without sensitive 6th Generation Human Intelligence Deployment we will not see the way to improve long-term profitability of the firm. And like it or not, profits have to be made – if only to ensure adequate funds to reinvest in new machines, telecommunications systems, and human resources.

HID in Asia is quite different in style and content from those processes operating in the USA or in Europe. In Asia there is often a rather open acceptance of opaqueness, and of gift-giving that seems in some cases close to bribery and nepotism. Indeed corruption, while actually illegal, is rife in this region. What the outsider has to realize is that individual relationships and the human side of management – at the personal level – is much more important than in the West. Of course, in the West we would never suggest that HID was not intimately linked with an individual's personal development – but in Asia such relationship management systems are a way of life. Intimate personalization may be aspirational in the Western HRM department at a policy level, enunciated to create calm, but with senior management knowing they may never operate in this way. In this book we are saying that insiders and outsiders must learn to work and make compromises with each other in a cooperative fashion, and not just *imply* that they are doing this. We understand the Asian 'rules' of *guanxi* (Chinese: gift-giving as acknowledgement of a connection) and *konne* (Japanese: derived from the English 'connection'), and how in choosing managers in the West we have to bear in mind all cultural antecedents and psychometric profiles of the actors in the alliances. Whatever we call it, it is the right *guanxi* that makes all the difference in ensuring that the business will be successful. Yet we must be crystal clear that relationships in Asia are not simply between companies but between individuals at a very personal level. We consider the deployment of Human Intelligence (HID) to be the co-evolution of management action supporting the training of individuals together with the development of cooperative attitudes expressed throughout the organization. Hence the

6th Generation – to progress beyond the HR-practices generally in use at present.

We consider the countries located in the Asian region to be poised for great growth in the future. In addition they will attract much foreign investment thereby implying the firms therein will need to move to new methods of joint management in the global stream of business. Thus we do not focus on traditional expatriate assignments to Asia but on HID as part of a global HR strategy. We have seen the US enter a dominant phase in its economic cycle while much of Asia still struggles to recover from its crisis. Even so, all nations and their peoples have become interlinked though trade with the spread of multinational firms throughout Asia, the Americas and Europe – they are truly global. As such they need to learn how to develop their HID capability for their own good, and through that they will also help develop their local staff. We do not propose a homogenization of the world's cultures; rather we believe that each nation can help make leisure and work a richer local experience – provided we learn to understand how and why people react as they do.

Over several decades we have seen increasing sophistication in the selection and deployment of the personnel in some firms. As the complexity of their operations became greater so their HRM departments acquired more tools in order to ply their trade. They grew from 'hire and fire' groups to ones who learned how to negotiate with and against the unions, and more recently they have learnt how to garner knowledge. They have found that it is beneficial for the firm to be able to instruct all staff in the 'how–what–when' of their processes and latterly the more important 'who'. The realization is that both the manufacturing and the service sector, each having suffered heavy downsizing over the past decades, now needs to be a knowledge-aware society. It is the HRM function that can engage all staff in this process through suitable selection, placement and training according to the needs of both the organization and the individuals concerned.

However, we find that these changes in the way that the HRM operates in different firms is not constant, nor in different countries are their methods necessarily similar. Although much research has taken place in firms located in the developed world, their understanding of the difficulties facing their multinational subsidiaries is weak. And weaker still is the knowledge about how to cope when the developed meets with the developing, notwithstanding the depth of history behind each firm. In fact this may indeed be an impediment – as we often see the brash, young American firm and its employees attempting to tell the workers in a developing country how to do things 'the way they should be done'. They are forgetting that some of these workers stem from very old societies with thousands of years of sophisticated social

HRM who uphold strongly held internalized rules of behaviour that brashness grates against.

This is an academic book – it is concerned with theories and practices developed in the field by authors very knowledgeable in Asian affairs. As such it gives strong pointers to the nature and role of the developing HID function in firms in Asia. The chapters offer a mobile exchange of perspectives on HID today; indeed they present this area not as a new field of specialization, but rather as a field of tension and experimentation. There is a sister book by Kidd, Li and Richter (*Advances in Human Resource Management in Asia*, 2001) which focuses initially on the practice of HRM in Asia (in Japan, Korea, and Southeast Asia), and then looks particularly on the experiences of Western firms in China. We suggest each book is a complement of the other.


## THE STRUCTURE OF THE BOOK

This book is surely not the last word in understanding the diverging spheres of culture and management in Asia, but it is intended to put our knowledge a decisive step forward. The chapters seek to represent a wide range of perspectives, as viewed by academics, consultants, and business people.

The book has eleven chapters and is organized in three parts. The chapters in Part I consider the strategic aspects of resource management in a global context; Part II reviews the deployment of human intelligence in Asia often from a socio-psychological perspective; and finally, in Part III, the authors consider how to avoid the 'us and them' scenario through the development of cultural literacy.

### Globalization, Strategic Management, and the Deployment of Human Intelligence

John Bratton (Chapter 2) writes on the integration of human intelligence and strategic management through the international strategic human resource management (HRM) function. He examines the concept of strategic management and its framework and then explores the links between business strategy and strategic-HRM. Just as the new HRM model is contested, so too is the notion of strategy. So, before looking at some of the issues associated with the strategy-HRM concept, the second part of the chapter considers the problems associated with the 'strategy' element of the term 'strategic HRM' (SHRM) and some issues associated with strategic HRM. The third part concentrates on the HRM-organization performance link and the presumption that workplace innovations associated with the new HRM model actually make a difference to organizational performance. This leads to the final section that examines international and comparative HRM issues, especially in the Asian community.

This chapter addresses a number of questions that the new SHRM paradigm raises. How do 'big' MNC (multinational corporation) corporate decisions impact on HRM? Do national models of HRM exist? Is it relevant to speak of an Asian HRM model? How transferable are national HRM models? A common theme running through this chapter, much of it from academe, points out that there are fundamental structural constraints that attest to the complexity of implementing the new HRM model where there are global product markets but local labour markets must be involved for the product's creation.

Ronald Jacobs (Chapter 3) reviews employee competence and human intelligence management in a global organization. He suggests that organizations need to manage their human competence and need to consider a taxonomy of employee development as a means to achieve this goal. The taxonomy presented describes a range of employee abilities, relative to the work that needs to be done. Specifically, he discusses how many performance improvement efforts have led to shifts in job responsibilities in several organizations. Indeed, these shifts in perspective form the primary rationale for proposing the taxonomy of employee development that is described here. Secondly, he proposes a taxonomy of employee development which emphasizes five categories of employee abilities. Finally, he discusses the implications of the taxonomy for HRM research and practice.

The overall goal is to contribute further to our understanding of how employee abilities relate to organizational performance. It is now a well-accepted fact that the most knowledgeable and skilful employees make a disproportionate (positive) contribution to organizational outcomes. Until recently, however, most research efforts have sought to establish the truth of this general principle. It now seems appropriate to move beyond that point to address further questions on this topic especially when involved in global strategies having local effects.

**Human Intelligence Deployment in Asia**

In Part II we have five chapters, two of which look broadly at American business culture and how it impinges on Asian culture, ritual and behaviour. The others look more directly at Asian behaviour and the need to raise consciousness.

William Czander and Dong Hwan Lee (Chapter 4) write on 'employee commitment' in a socio-psychological sense while comparing Asian and

American business and culture. They say that institutional transference is the key to commitment. In this chapter they review the literature on commitment to an organization and suggest that 'commitment' offers the greatest potential for self system integration, group cohesion, and attachments to fellow employees and the organization. Against this positive aspect it is suggested that it is the organization's management who foster fragmentation of commitment by unconsciously undermining the employee's motivation to be committed, and thus frustrating the employee's unconscious wish for a maternal haven. It is suggested that all employees seek in their tasks and relationships a kind of spiritual involvement, derived from their real relationships in their family group, or from their romance with their fantasized 'family' wherein the workplace organization acts as a pseudo-family. A near accomplishment or even encouragement of the belief that these wishes and fantasies may be attained will precipitate a powerful positive transference and genuine commitment.

Finally, based upon case-studies, it is seen that organizational leaders more often than not undermine the creation of a positive transference, and in its place promote a false or disingenuous commitment.

Oliver Tzeng (Chapter 5) presents a 'psychosomantic model of social-economic behaviours' and anchors this in an Asian context. This chapter presents a multidisciplinary approach to the study of business behaviour in the context of cross-cultural (Chinese–Western) interactions across the Pacific Ocean. Under this approach, business behaviours are evaluated with reference to various substantive components of a theoretical process-oriented framework structured under the Psychosemantic Process Model of Human Economic Behaviours. Here we suggest such economic behaviour covers a broad range of transactions – from the simple consumer purchase of a household good in the American market imported from China, to a complex long-term investment in China by an American public or private representative.

All behaviour concerns simultaneously multidisciplinary issues in macro- and micro-economic considerations, psychological attitudes, cognitive decisions, legal standards, social values, etc. Therefore, any single economic act can be characterized in terms of the distinct components of the proposed theoretical model. He discusses their impact with respect to the different legal foundations and standards in business transactions – from contract formation, industrial manufacturing, product distribution, and allocation of profits or losses, to conflict resolution. At the conclusion, this chapter offers a set of principles and strategies for effective communications between the Chinese and foreign businesspersons who may posses little knowledge of between-cultural disparities.

Cherlyn Granrose (Chapter 6) reviews 'Confucianism' and how this widespread Asian 'rule-set' might affect American career assumptions. This chapter examines 'The Analects of Confucius' to identify the assumptions behind this perspective. Following on, it examines the 'boundary-less careers' literature to critique the assumptions common to North American careers research. From the clarification that becomes possible we are able to generate propositions about international, especially Asian-inclined, careers that in some ways complement and in some ways differ from traditional Western ways of exploring careers. Propositions are offered for generating hypotheses that might be useful in conducting further career research in Asia; and indeed, in suggesting new avenues for career research in the US.

Dennis Heaton and Harald Harung (Chapter 7) present 'Awakening Creative Intelligence for Peak Performance: Reviving an Asian Tradition'. Whereas the objective approach of modern science has been prevalent in the West, Eastern traditions have maintained subjective paths for developing enlightened wisdom. Nevertheless current global scientific progress tends towards the imprinting of Western technologies and Western 'culture' to all parts of the world. They suggest that balancing this Western influence with Eastern approaches to inner development can offset the growing risk that Western science is producing imbalance in the systems of nature and alienating the roots of indigenous cultures.

Meditation practices from Eastern traditions have been adopted for human resource development in European and American companies as well as in Asian companies. Research on the Transcendental Meditation programme has found that reduced stress, reduced health care costs, improved psychological well-being, and more holistic thinking has come about through the adoption of TM programmes. The growth of wisdom through such practices has been found to be not just a phenomenon of Eastern cultures, but an inherent and universal human potential. It is suggested that management education and development programmes throughout the world ought to integrate the best of Western knowledge with Asia's contribution of awakening inner intelligence. This chapter suggests ways towards this enlightenment.

Mohan Raj Gurubatham (Chapter 8) writes on consciousness as a base for cognition. This chapter reviews many 'romantic' analyses of popular human resource notions now popularized as 'collectivism versus individualism', and it explains the striking discrepancies that many Western expatriates have encountered during the pre-crises years in human resources and generally in human behaviour in East Asian countries. Many of these discrepancies can be explained within the broader framework of Hofstede's or Bond's writings, as well as those of Edward Hall. Herein, new directions grounded in fundamental roots of Western and Eastern modes of consciousness are explored for development in the new century after the current effects of the industrial revolution have dissipated.

These notions disabuse the popular Western expatriate belief that Asians do not manage process in organizational work practices, such as enacting transparency. An explanation of 'collectivism' is then embedded within Hall's thesis of high context cultures. This suggests that in East Asia collectivism lies within the high context dimension, so these persons manage processes informally through broad modes of human information processing (except in Singapore where both the overseas Chinese and 'original' inhabitant groups demonstrate an appalling lack of transparency).

**Making Sense of the 'Us and Them' Scenario**

The chapters in Part III explore the Western view of managing human resources in Asia, and how it is emerging into a more sophisticated 'view of the other' rather than projecting so strongly its ethnocentric past upon an unsuspecting and sometimes unsophisticated audience. In England, the old adage 'Go West, young man' offered the promise of a better world, one of riches, and so on. But now, Americans hearing the same phrase have to be more cautious, as those people 'out there in the West' are not of English or of European stock, and do not hold the same broad principles. 'Out there' is remarkably different! Indeed, we have to make sense of the 'us and them' scenario to better the deployment of human resources in Asia.

Lionel Stapley (Chapter 9) gives us 'Beyond Culture Shock: Developing Cross-Cultural Awareness'. This chapter aims to provide the reader with a way of thinking about the effects of societal culture on behaviour and about developing cross-cultural awareness that will result in more productive working relationships. It begins with a clarification of what is meant by culture, using work previously developed by Stapley (1996), and uses this to provide an explanation for the way that diverse cultures develop. He then explores the notions of 'organizational socialization' and 'culture shock' as a means of explaining what happens when we operate in an alien societal culture. This leads, in the latter part of the chapter, to the development of a method of achieving cross-cultural awareness and understanding that will create effective relationships.

Allan Bird and Joyce Osland (Chapter 10) recognize the American dilemma – of projecting too much from an internal perspective. They present their views on whether Westerners can make sense of Asian cultural paradoxes, acknowledging that much cross-cultural training and research occurs within the framework of bipolar cultural dimensions. While this 'sophisticated stereotyping' is helpful to a certain degree, it does not convey the complexity found within cultures. People working across cultures are frequently surprised by cultural paradoxes that do not seem to fit the descriptions they have learned. The authors identify the sources of cultural paradoxes and introduce the idea of value trumping – that in a specific context, certain cultural values take precedence over others. Thus, culture is embedded in the context and cannot be understood fully without taking context into consideration.

To decipher cultural paradoxes, the authors propose a model of cultural sense-making, linking schemas to contexts. They spell out the implications of this model for those who teach culture, for people working across cultures, and for multinational corporations.

Philip Merry (Chapter 11) considers 'Cultural Literacy – Its Link to Business Success in Asia-Pacific'. He says many MNCs operating across borders get into difficulty by not recognizing the vital connection between the so-called 'soft issues' of emotion and culture, and the 'harder issue' of business profitability. This chapter shares experiences gained by the author in Asia-Pacific whilst working with a variety of major MNCs where inter-cultural and emotional relationship issues were the main focus. Essentially it is supported by three broad models: (a) a 4-step model for developing 'Cultural Literacy', (b) a model for developing 'Emotional Literacy', and (c) a 7-dimension model of culture taken from Trompenaars (1993). These models of cultural and emotional literacy are grounded in the author's extensive Asian consulting experience.


## CONCLUSIONS

The theories and the practices so well expressed above lead us to suggest that the Asian region will help the world develop richer values in society and in the workplace. So in summing up the issues of the different sections and chapters, we come to the following conclusions:

- Multinationals have to develop a new approach towards expatriate assignments, focusing more on a holistic approach to deploy human intelligence than the former one-dimensional HR-strategies have been able to provide.
- Asia is very complex and it represents a dynamic laboratory in which to deploy human intelligence. One must be sophisticated, and be culturally educated in order to deploy resources well.

- It is not sufficient to classify and to use the Western cultural stereotypes to assess the rich and uncommon potential of Asian human resources.

There was a tendency for European and American managers to venture abroad carrying with them restricted stereotypes of their intended hosts – even today this still applies ('We want their market, they want our high technology' can often be heard). It has become recognized by the Japanese, in their outward venturing and in their creation of manufacturing alliances, that the nations representing 'Europe' are quite diverse. So managers from Europe and from America should not consider Asia, in its vastness, as a homogeneous region: people cannot be classified simply as 'white, black or yellow'. Even China with its singular national language has recognized that it supports many ethnic subgroups (with their own verbal languages). It is not surprising therefore that managers from Singapore, or from Hong Kong, have their own particular difficulties in promoting business in mainland China. So how may an outsider do better?

We are very confident that the views expressed in this book will give guidance on the acquisition of insights into Asian ways, and to the development of a consciousness which may be accepting of this way of working. Through the HID and through being more culturally aware we should be able to better enjoin in commercial and non-commercial ventures between Western and Eastern enterprises.

## References


Backman, M. (1999) *Asian Eclipse. Exposing the Dark Side of Business in Asia* (Singapore: Wiley).

Détienne, M. and Vernant, J. P. (1978) *Cunning Intelligence in Greek Culture and Society*, trans. J. Lloyd (Hassocks: Harvester Press).

From, J. and Holmgren, C. (2000) 'On the "Chineseness" in Chinese Educational Practices', *Nordic Institute of Asian Studies Newsletter – NIASnytt*, (3), pp. 20–1.

Gilpin, R. (2000) *The Challenge of Global Capitalism. The World Economy in the 21st Century* (Princeton, NJ: Princeton University Press).

Haley, G. T. and Haley, U. C. V. (2000) 'Boxing with Shadows: Competing Effectively with the Overseas Chinese and Overseas Indian Business Networks in the Asian Arena', in U. C. V. Haley (ed.), *Strategic Management in the Asia Pacific. Harnessing Regional and Organizational Change for Competitive Advantage* (Oxford: Butterworth Heinemann).

Jaques, E. (1989) *Requisite Organisation* (London: Casson Hall).

Kanter, R. M. (1995) 'Thriving locally in the Global Environment', *Harvard Business Review*, Sep/Oct, pp. 151–61.

Kidd, J. B. and Richter, F.-J. (2001) 'The Hollowing Out of the Workforce: A Potential for Organizational Learning', *Human Systems Management* (in press).

Kidd, J. B., Li, X. and Richter, F.-J. (2001) *Advances in Human Resource Management in Asia* (London and New York: Palgrave).

Pucik, V. (1992) 'Globalisation and Human Resource Management', in V. Pucik, N. Tichy, and C. Barnett (eds), *Globalising Management: Creating and Leading the Competitive Organisation* (New York: Wiley).

Raphals, L. (1992) *Knowing Words: Wisdom and Cunning in the Classical Traditions of China and Greece* (Ithaca, NY: Cornell University Press).

Reich, R. (1991) *The Work of Nations: Preparing Ourselves for Twenty-First-Century Capitalism* (New York: Alfred A. Knopf).

Richter, F.-J. (1999) *Strategic Networks. The Art of Japanese Interfirm Cooperation* (New York: International Business Press).

Richter, F.-J. (2001) *Confucius' New Clothes: Re-designing the Texture of Asian Business* (forthcoming).

Stapley, L. F. (1996) *The Personality of the Organisation: A Psycho-Dynamic Explanation of Culture and Change* (London: Free Association Books).

Sun Tzi (1996) *The Art of War*, trans. and retold by Stefan Rudnicki (Los Angeles: Newstar Press).

Takashima, T. (1985) *Kotowaza no izumi* [Fountain of Japanese Proverbs] (Tokyo: Kodansha).

Trompenaars, F. (1993) *Riding the Waves of Culture: Understanding Cultural Diversity in Business* (London: The Economist Books).

# 2 International Strategic Human Resource Management: Integrating Human Intelligence and Strategic Management

John Bratton

## INTRODUCTION

In mainstream business literature, globalization is usually described in terms of shifts in traditional patterns of international investment, manufacturing, and trade (Dicken, 1992). Despite the trend towards globalization, at the beginning of the millennium, the Asian Region still exhibits enormous economic, political and cultural diversity (Moore and Devereaux Jennings, 1995). Competitive advantage in some Asian countries is predicated on cheap unskilled labour, yet others use both unskilled labour and highly skilled workers. Some countries are politically stable, while others are subject to instability. Some countries are marked by cultural homogeneity, while others are marked by cultural diversity. The countries located in the Asian Region speak at least seven different major languages and believe in widely different religions and philosophies, ranging from Buddhism and Hinduism to Islam and Christianity (Harris and Moran, 1991). Thus, Robertson (1995) has asserted: 'one paradoxical consequence of the process of globalization . . . is not to produce homogeneity but to familiarize us with the greater diversity, the extensive range of *local cultures* (my emphasis, and cited by Parker, 1999, p. 235). The existence of strong local cultures means that multinational corporations are increasingly dependent upon multicultural workforces to retain their competitive advantage. It is not surprising that the deployment of human intelligence and HRM systems in general in the Asian Region will differ in many ways. But, how 'imitable' are Western strategic human resources management (SHRM) models – based on Western business practices, culture and values – to Asian business systems, culture and values?

In this chapter we explore various strategic issues associated with HRM which will help practitioners identify important imitable and inimitable elements of the Western HRM paradigm. This chapter first examines strategic management concepts and explores the links between business strategy and HRM. The second part of the chapter considers the problems associated with the 'strategy' element of the term 'strategic HRM' (SHRM). The third part examines some international and comparative HRM issues relevant to corporations operating in the Asian Region. The chapter addresses a number of questions, some essential to our understanding of how post-industrial organizations work, which the new SHRM paradigm raises. How do 'big' corporate decisions impact on HRM? Does the evidence suggest that firms adopting a 'strategic' HRM approach experience superior performance? Do national models of HRM exist? How transferable are national HRM models? There is a common theme running through this chapter, much of it from academe, which points out that there are fundamental structural constraints that attest to the complexity of implementing the new HRM model where there are global uniform product markets but local diverse labour markets.

## STRATEGIC MANAGEMENT

The word 'strategy' was first used in English in 1656 and comes from the Greek noun 'strategus', meaning 'commander in chief'. The development and usage of the word suggests that it is composed of *stratos* (army) and *agein* (to lead) and in its military context means 'to produce large-scale operations' (Aktouf, 1996, p. 93). In a management context, the word 'strategy' has now replaced the more traditional term, long-term planning, to denote an activity that top managers perform, in order to accomplish an organization's goals. Wheelen and Hunger (1995, p. 3) define strategic management as 'that set of managerial decisions and actions that determines the long-run performance of a corporation'. Aktouf (1996) takes a similar view when he sees strategy as the maintenance of a 'vision of the future' that is constantly updated by data on both the internal and the external environment. Other definitions place less emphasis on conscious 'plans' on the part of senior management, but prefer instead to focus on a stream or 'pattern' of activity over time to achieve performance goals (Mintzberg, 1994; Hill and Jones, 1998; Watson, 1999). As Hill and Jones (1998, pp. 3–4) have argued: 'A strategy is a specific pattern of decisions and actions that managers take to achieve an organization's goals . . . For most if not all organizations, an overriding goal is to achieve superior performance . . . [Therefore] a strategy can often be defined more precisely as the *specific pattern of decisions* and actions that managers take to achieve superior organizational

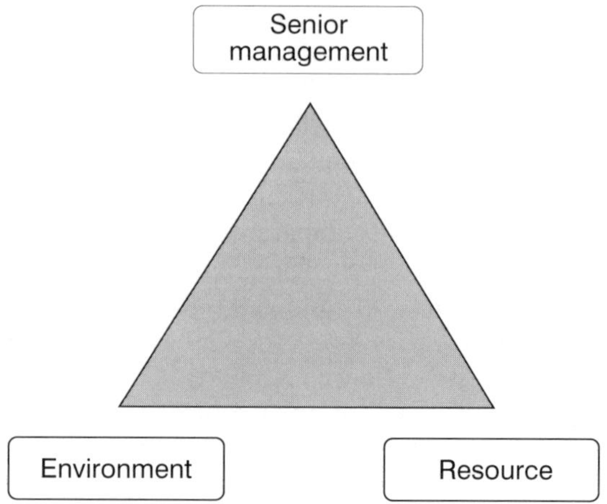



*Figure* 2.1   The three traditional poles of a strategic plan
*Source*:   Adapted from Aktouf, 1996.

performance.' Of this school, Watson (1999) has similarly argued that 'the strategy of an organization is thus *not the plan that the organization follows* but a *pattern that unfolds over time* in which formal planning can be found to occur to a greater or lesser degree' (1999, p. 23). Strategic management therefore is considered a continuous activity, undertaken by the upper echelon of the organization, that requires constant adjustment of three major interdependent poles: the values of senior management, the environment, and the resources available (see Figure 2.1). Strategic management emphasizes the necessity to monitor and evaluate environmental opportunities and threats in the light of an organization's strengths and weaknesses. Hence, any changes in the environment and the internal and external resources must be monitored closely so that the goals pursued can, if necessary, be adjusted. The goals should be flexible and open to amendment subject to the demands and constraints of the environment and what takes place in the status of the resources.

## MODEL OF STRATEGIC MANAGEMENT

In the descriptive and prescriptive management texts, strategic management appears as a cycle in which several events follow and feed upon one another. The strategic management process is typically broken down into five events or steps: (1) organization's direction, (2) environmental analysis, (3) strategy formulation, (4) strategy implementation, and (5) strategy evaluation. At the corporate level, the strategic management process includes activities that range from appraising the organization's current mission and goals to strategic evaluation.

The first step in the strategic management model begins with senior managers evaluating their position in relation to the organization's current *mission and goals*. The mission describes the organization's values and aspirations. It is the organization's *raison d'être*, and indicates the direction senior management is going. A goal is a desired future state that the organization attempts to realize (Daft, 1998, p. 42). *Environmental analysis* looks at the internal organizational strengths and weaknesses and the external environment for opportunities and threats. The factors that are most important to the organization's future are referred to as strategic factors and are summarized with the acronym SWOT, meaning Strengths, Weaknesses, Opportunities, and Threats. *Strategic formulation* involves senior managers evaluating the interaction of strategic factors and making *strategic choices* that guide the organization to meet its goal(s). Some strategies are formulated at the corporate, business, and specific functional level such as marketing and HRM. The use of the term 'strategic choice' raises the question of who makes decisions in work organizations and why they are made (McLoughlin and Clark, 1988). The notion of strategic choice also draws attention to strategic management as a 'political process' whereby strategic choices on issues such as resources are taken by a 'power-dominant' group of senior managers within the organization (Child, 1972; Purcell and Ahlstrand, 1994). The strategic choice perspective on organizational decision-making makes the discourse on strategy 'more concrete'; it also provides important insights into how the employment relationship is managed.

*Strategy implementation* is an area of activity that focuses on the techniques used by managers to implement their strategies. In particular, it refers to activities that deal with leadership style that is compatible with the strategies, the structure of the organization, the information and control systems, and the management of human resources. Leading management consultants and academics (see Champy, 1996; Kotter, 1996) emphasize strongly that leadership is the most important and difficult part of the strategic implementation process. *Strategy evaluation* is an activity in the strategic management process that determines to what extent actual change and performance matches desired change and performance. The strategic management model depicts the five main activities undertaken by senior managers as a rational and linear process. However, it is important to note that it is a *normative* model, that is, it shows how strategic management *should* be done and hence *influencing* managerial processes and practices, rather than describes what is actually done by senior managers (Wheelen and Hunger, 1995). As we have already noted, the notion that strategic decision-making is a political process implies a potential gap between the theoretical model and reality.

## HIERARCHY OF STRATEGY

Another aspect of strategic management in the multidivisional business organization concerns the organizational level to which strategic issues apply. Conventional wisdom identifies different levels of strategy: (1) corporate, (2) business, and (3) functional (see Figure 2.2). These three levels of strategy form a hierarchy of strategy within a large corporation. In different companies the specific operation of the hierarchy of strategy might vary between 'top-down' and 'bottom-up' strategic planning. The top-down approach resembles a 'cascade', whereas the 'downstream' strategic decisions are dependent on higher 'upstream' strategic decisions (Wheelen and Hunger,1995).

*Corporate-level strategy* describes a corporation's overall direction in terms of its general philosophy towards growth and the management of its various business units. Such strategies determine the type of businesses a corporation wants to be in and what business units should be acquired, modified or sold. This strategy addresses the question *What business are we in?* Devising a strategy for a multidivisional company involves at least four types of initiatives:

- Establishing investment priorities and steering corporate resources into the most attractive business units.
- Initiating actions to improve the combined performance of those business units that the corporation first got into.
- Finding ways to improve the synergy amongst related business units in order to increase performance.
- Decisions dealing with diversification.

*Business-level strategy* deals with decisions and actions pertaining to each business unit. The main objective of a business-level strategy is to make the unit more competitive in its marketplace. This level of strategy addresses the question *How do we compete?* Although business-level strategy is guided by 'upstream' corporate-level strategy, business unit management must craft a strategy that is appropriate for their own operating situation. Porter (1980) made a significant contribution to our understanding of business strategy by formulating a framework that describes three competitive strategies: low-cost leadership strategy, differentiation strategy, and focus strategy. The low-cost leadership strategy attempts to increase the organization's market share by emphasizing low unit cost compared with competitors. In a differentiation competitive strategy, managers try to distinguish their services and products – such as brand image or quality – from others in the industry. With the focus competitive strategy, managers focus on a specific buyer group or regional market.

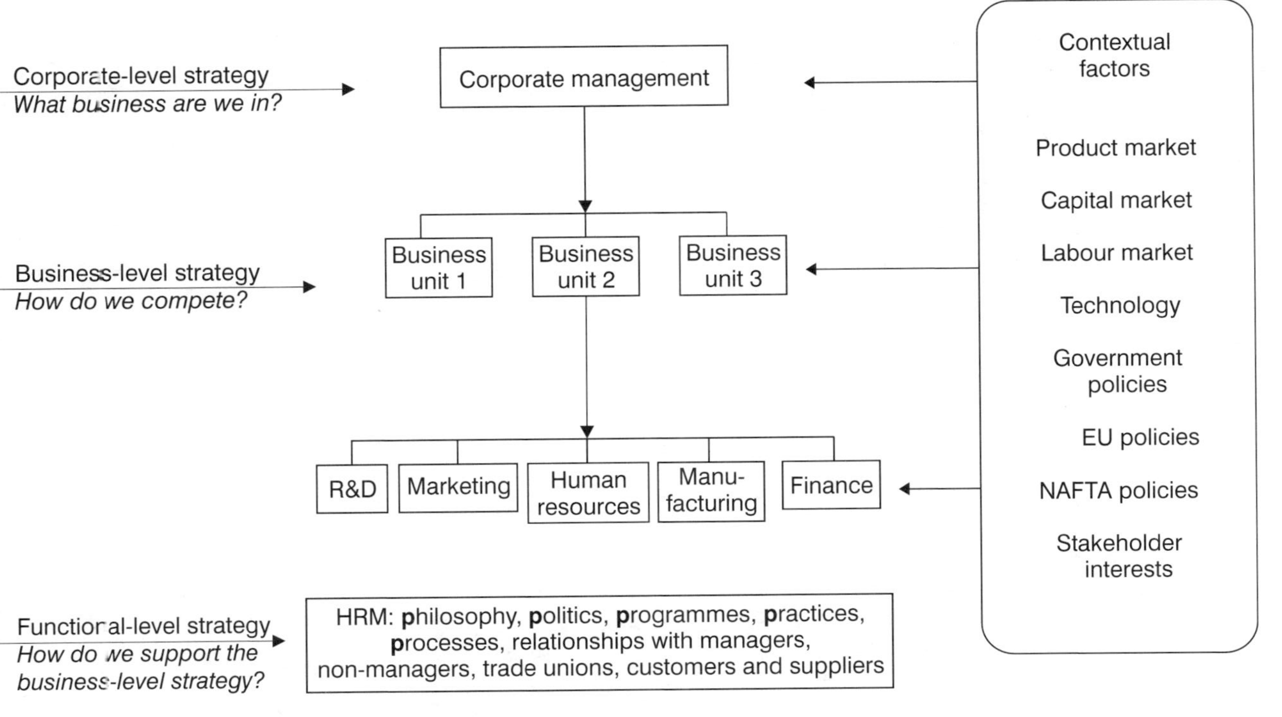


Figure 2.2   Hierarchy of strategic decision-making
Source:   Adapted from Hierarchy of Strategic Decision-Making in Bratton and Gold (1999).

Miles and Snow (1984) also made an important contribution to the strategic management literature. These authors identified four modes of strategic orientations: defenders, prospectors, analysers, and reactors. *Defenders* are companies with a limited product line and the management focus on improving the efficiency of their existing operations. Commitment to this cost orientation makes senior managers unlikely to innovate in new areas. *Prospectors* are companies with fairly broad product lines that focus on product innovation and market opportunities. This sales orientation makes senior managers emphasize 'creativity over efficiency'. *Analysers* are companies that operate in at least two different product market areas, one stable and one variable. In this situation senior managers emphasize efficiency in the stable areas and innovation in the variable areas. *Reactors* are companies that lack a consistent strategy–structure–culture relationship. Thus, in this reactive orientation senior management's responses to environmental changes and pressures tend to be piecemeal strategic adjustments. According to Miles and Snow, competing companies within a single industry can choose any one of these four modes or types of strategies and adopt a corresponding combination of structure, culture, and processes consistent with that strategy in response to the environment. These strategic choices help explain why companies facing similar environmental threats or opportunities behave differently and why they continue to do so over a long period of time (Wheelen and Hunger, 1995). In turn, the different competitive or business strategies influence the 'downstream' functional strategies.

*Functional-level strategy* pertains to the major functional operations within the business unit; including research and development, marketing, manufacturing, finance, and human resources. Typically, this strategy level is primarily concerned with maximizing resource productivity and addresses the question *How do we support the business-level competitive strategy?* The three levels of strategy – corporate, business, and functional – form a hierarchy of strategy within a large multi-divisional corporation. Strategic management literature emphasizes that the strategies at different levels must be fully integrated (Wheelen and Hunger, 1995).

## BUSINESS-LEVEL STRATEGY AND HRM

Strategic management texts emphasize that each level of strategy forms the strategic environment of the next level in the corporation. At the functional-level, HRM strategy is implemented to facilitate the business strategy goals. An HR strategy can be defined as 'The pattern that emerges from a stream of important decisions about the management of human resources, especially those decisions that indicate management's major goals and the means that are (or will be) used to pursue them' (Dyer, 1984, p. 159). The nature of the links between HRM strategy and business strategy has received much attention in the literature. A range of business-HRM links have been identified and classified in terms of a proactive–reactive continuum (Kydd and Oppenheim, 1990), and in terms of environment – human resource strategy – business strategy linkages (Bamberger and Phillips, 1991). In the 'proactive' orientation, the HRM professional has a seat at the strategic table and he or she is actively engaged in strategy formulation. In Figure 2.2, this type of proactive model is depicted by the two-way arrows on the left-hand side showing both downward and *upward* influence on strategy. At the other end of the continuum is the 'reactive' orientation which sees the HRM function as fully subservient to corporate and business-level strategy, and corporate and business-level strategies ultimately determining HRM policies and practices. Once the business strategy is determined, without the involvement of the HRM professional, HRM policies and practices are implemented to support the chosen competitive strategy. This type of reactive orientation would be depicted in Figure 2.2 by a *one-way downwards* arrow from business to functional-level strategy. In this sense the practice of strategic HRM is concerned with the challenge of matching the philosophy, policies, programmes, practices and process, the 'five P's', in a way which will stimulate and reinforce different employee role behaviours appropriate for each competitive strategy (Schuler, 1989).

The importance of the environment as a determinant of HRM policies and practices has been incorporated into some models. Extending strategic management concepts, Bamberger and Phillips' (1991) model depicts links between three poles: the environment, human resource strategy and the business strategy. In the hierarchy of strategic decision-making model the HRM strategy is influenced by contextual variables such as markets, technology, national government policies, EU policies and trade unions. Purcell and Ahlstrand (1994) argue that those models that incorporate contextual influences as a mediating variable of HRM policies and practices tend to lack 'precision and detail' into the precise nature of the environment linkages and 'much of the work on the linkages has been developed at an abstract and highly generalized level' (p. 36).

In the late 1980s, Purcell made a significant contribution to research on business-HRM strategy. Drawing on the literature on 'strategic choice' in industrial relations (e.g. Thurley and Wood, 1983; Kochan, Katz and McKersie, 1986) and using the notion of a hierarchy of strategy, Purcell (1989) identifies what he labels 'upstream' and 'downstream' types of strategic decisions. 'Upstream' or 'first-order' strategic decisions are concerned with the long-term direction of the corporation. If a first-order decision is made to take over another enterprise, for example, a Japanese company acquiring a water company in China, a second set of considerations apply concerning the extent to which the new operation is to be integrated with or separated from existing operations. These are classified as 'down-stream' or 'second-order' strategic decisions. Different HRM approaches are called 'third-order' strategic decisions because they establish the basic parameters of labour management in the workplace. In theory, wrote Purcell, 'strategy in human resources management is determined in the context of first-order, long-run decisions on the direction and scope of the firm's activities and purpose...and second-order decisions on the structure of the firm' (1989, p. 71). In a major study of HRM in multi-divisional companies Purcell and Ahlstrand (1994) argue that what actually determines human resource management policies and practices will be determined by decisions at all three levels, and by the ability and leadership style of local managers to follow through goals in the context of specific environmental conditions. Much of the strategic human resource management (SHRM) literature has focused on two aspects of the strategy debate: the integration or '*fit*' of HRM strategy with business strategy and the '*resource-based*' model of strategic HRM. The next section takes a critical look at these influential SHRM models.


## STRATEGIC HUMAN RESOURCE MANAGEMENT

Although the roots of the strategic literature on HRM are in 'manpower' [*sic*] planning, it is the normative HRM models developed in the 1980s that made the strategic concept central to research productivity in this area (Cappelli and Singh, 1992). In the 1980s, scholars attached the prefix 'strategy' to the term 'human resource management' and the notion of 'strategic integration' became prominent in the HRM literature. Interest among practitioners of linking the strategy concept to HRM can be explained by the pressure to enhance the status of HRM professionals within companies (Purcell and Ahlstrand, 1994) at a time when 're-engineering' is questioning the need for HRM specialists in a 'flatter' organizational structure.

One key feature of Beer *et al.*'s (1984), model of HRM is '*strategic integration*'; in particular the need to establish a close two-way relationship or 'fit' between the external business strategy and the elements of the internal HR strategy. The concept of integration has three aspects: the integration or 'cohesion' of HR policies and practices in order to complement each other and to help achieve strategic goals; the internalization of the importance of

HR on the part of line managers; and third, the integration of all workers into the business to foster commitment or an 'identity of interest' with their organization. The basic proposition developed here is that if these forms of integration are implemented, workers will be more cooperative, flexible and willing to accept change, and therefore the organization's strategic plans are likely to be more successfully implemented. In this section we examine the nature of the relationship of one element of integration: the strategic planning–HRM link. This approach to strategic HRM is referred to as the 'Matching' model. We also examine an alternative view of strategic HRM: the 'Resource-based' model.


## THE MATCHING SHRM MODEL

The underlying premise of this influential model is that high-wage countries in the Western hemisphere can only gain competitive advantage through adopting Porter's (1980; 1985) generic 'low cost' or 'differentiation' strategy. Further, each Porterian competitive strategy involves a unique set of responses from workers or 'needed role behaviours' and a particular HRM strategy that might generate and reinforce a unique pattern of behaviour (Schuler and Jackson, 1987; Cappelli and Singh, 1992). Thus, the practice of strategic HRM is concerned with the challenge of matching the philosophy, policies, programmes, practices and process, the 'five P's' (see Figure 2.2), in a way which will stimulate and reinforce different employee role behaviours appropriate for each competitive strategy (Schuler, 1989). Similarly, each type of Miles and Snow's (1984) competitive strategies – 'defender', 'prospector' and 'analyser' – will require that an organization's HRM policies and practices should be configured and managed in a way that is congruent with each particular strategy.

The publication of Fombrun *et al.*'s *Strategic Human Resource Management* (1984) generated early interest in the 'matching' model. It was argued that 'HR systems and organizational structure should be managed in a way that is congruent with organizational strategy' (p. 37). This is similar to Chandler's (1962) distinction between strategy and structure and his often quoted maxim that structure follows strategy. In the Devanna *et al.* (1984) model, human resource management–strategy–structure follow and feed upon one another and are influenced by environmental forces (Figure 2.3). This basic model constituted the 'bare bones of a theory' on SHRM (Boxall, 1992).

The notion of 'fit' between an external competitive strategy and the internal HRM strategy is a central tenet of the HRM model advanced by Beer *et al.* The authors emphasize the analysis of the linkages between the

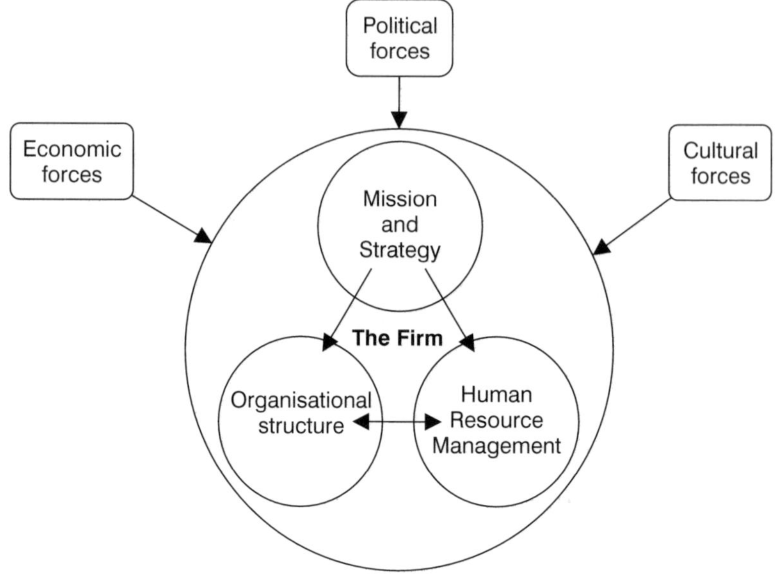



*Figure* 2.3    Matching model of strategic HRM
*Source*:    Adapted from Devanna *et al.*'s (1984) matching model of strategic HRM.

two strategies and how each strategy provides goals and constraints for the other. There must be a 'fit between competitive strategy and internal HRM strategy and a fit among the elements of the HRM strategy' (p. 13). There is some theorization of the link between product markets, organizational design and approaches to labour management. So, for example, a firm manufacturing large-batch products in a market where low cost is critical will, Beer *et al.* argue, need to develop a different approach to managing its workforce, than a firm manufacturing small-batch, customized products where quality is a key success factor. In the matching model, the relationship between business strategy and HRM strategy is said to be 'reactive' in the sense that HRM strategy is subservient to 'product market logic' and the corporate strategy. The latter is assumed to be the independent variable (Boxall, 1992, 1996; Purcell and Ahlstrand, 1994). As Miller (1987) emphasizes: 'HRM cannot be conceptualized as a stand-alone corporate issue. Strategically speaking it must flow from and be dependent upon the organization's (market oriented) corporate strategy' (cited in Boxall, 1992, p. 66).

**Limitations with the Matching Model**

The matching model can be critiqued on both conceptual and empirical grounds. On the first area, the conceptual, the matching model is predicted upon the rational view of strategic decision-making grounded in clearly definable predispositions and acts of planning, choice and action. That is, the third-order or 'internal' strategy – a carefully planned approach to how people at work are to be deployed, developed, motivated and controlled – is derived from first-order or 'external' strategy – a chosen approach of competing in the marketplace. In other words, it assumes that organizational controllers act rationally. But as many critical organizational theorist attest, strategic decisions are not necessary based on the output of rational calculation. The assumption that business-level strategy and HRM strategy have such a logical linear relationship is questionable given Whittington's (1993) work on strategy. As Legge (1995) emphasizes, the 'act of *consciously* matching HRM policy to business strategy is only relevant if one adopts the rationalistic "classical" perspective on strategy' (p. 103). The 'decision process' model and the 'political process' model appear to provide a more fruitful approach into strategic decision-making. Proponents of the two models argue that managerial rationality is limited by lack of information, time and 'cognitive capacity'. Rather than viewing strategic choices as the outcome of rational decision-making, Johnson (1987) opines that 'strategic decisions are characterized by the political hurly-burly of organizational life' (cited in Purcell, 1989, p. 72).

A second problem is with the prescriptive validity of the model. Some HRM theorists have questioned whether the fit metaphor is necessarily a desirable goal to achieve. In periods of market turbulence and financial stringency there is a tendency for corporate management to improve profitability by downsizing, decentralizing decision-making and applying more demanding performance outcomes at the unit level. As Purcell (1989, 1995) argues, this, in turn, encourages similar trends in HRM and industrial relations strategies. A multi-divisional company pursuing a strategy of acquisition, asset stripping, and downsizing might 'logically' adopt a set of HRM strategy that includes compulsory lay-off and a compensation system based on short-term performance results. In such a case, the business strategy and HRM strategy might 'fit', but as Legge points out, these HRM policies 'although consistent with such a business strategy, are unlikely to generate employee commitment' (1995, p. 126). In other words, achieving the goal of 'close fit' of business and HRM strategy can contradict the core 'soft' HRM goals of commitment, flexibility and quality.

Another problem centres on underlying structural variables in profit-driven economies, which seriously undermine the notion of strategic integration. Purcell's (1995) work, for example, demonstrates how the imperatives of the market place and 'rational' managerial decisions limits the adoption of the matching paradigm. He argues that when a financial-control mode of management and short-term investment criterion dominates, it tends to drive out long-term HR investment at the workplace and 'destroy' the basis of HRM as part of corporate strategy. He also notes that multi-divisional corporations are not monolithic; a range of possible patterns of corporate strategy is possible. It is, however, worth noting that the companies adopting a financial-control model, 'substantially out-perform the industry average' and, consequently, are viewed favourably by the capital markets. The new millennium continues to be marked by short-term financial expediencies in the Anglo-North American capital markets and this trend makes the adoption of non-economic and intangible values characteristic of the 'soft' HRM paradigm as part of a corporate strategy improbable.

On the second area for debate, the empirical, there are two related hypotheses: the first asks whether HRM strategies are in fact related to business strategy; the second question asks whether organizations that manage to achieve a 'tight fit' actually experience superior performance. The second and more challenging question, we address later in this chapter. As far as the first empirical question is concerned, it is evident that both survey-based research and case studies have generated only limited empirical support for the matching paradigm (Boxall, 1992, 1996; Legge, 1995). Purcell and Ahlstrand's (1994) study of multi-divisional companies found that HRM issues are 'rarely taken into account' in the formulation of corporate strategies. Downie and Coates's (1994) survey study of Canadian firms reported that HRM is taking on somewhat 'more strategic importance', but provided little evidence to substantiate the notion of strategic integration. Peck's (1994) survey study of the relationships between 'strategy, HR policies and the employment relationship' in 45 American firms concluded that the relationships between the three are 'more complex than previously assumed' (1994, p. 729). As a review of the literature makes clear, the fit metaphor has proven to be both conceptually and empirically elusive. We turn now to the second approach to strategic HRM, the 'Resource-based' model.


## THE RESOURCE-BASED MODEL

The resource-based model of SHRM draws attention to the strategic value of workplace knowledge and learning. The genesis of the 'resource-based' model can be traced back to Selznick (1957) who suggested that work organizations each possess 'distinctive competence' that enables them to outperform their competitors, and to Penrose (1959) who conceptualized the firm as a 'collection of productive resources'. More recently, Barney (1991) has posited that '*sustained* competitive advantage' (my emphasis) is not achieved through an analysis of its external market position but through a careful analysis of the firm's skills and capabilities; characteristics which competitors find themselves unable to imitate. Putting it in terms of simple SWOT analysis: the matching model emphasized the strategic significance of external 'Opportunities' and 'Threats'; and the resource-based perspective emphasizes the strategic importance of internal 'Strengths' and 'Weaknesses'. This model suggests that work organizations achieve sustainable competitive advantages by 'implementing strategies that exploit their internal strengths, through responding to environmental opportunities, while neutralising external threats and avoiding internal weaknesses' (Barney, 1991, p. 99).

Barney argues that four characteristics of resources and capabilities are important in sustaining competitive advantage: value, rarity, inimitability and non-substitutability. From this perspective, the collective learning in the workplace by managers and non-managers, especially how to coordinate workers' diverse knowledge and skills and integrate diverse information technology, is a strategic asset that rivals find difficult to replicate. Amit and Shoemaker (1993) make a similar point when they emphasize the strategic importance for managers to identify, *ex ante*, and marshal 'a set of complementary and specialized resources and capabilities which are scarce, durable, not easily traded, and difficult to imitate' to enable the company to earn 'economic rent' (profits). Thus, according to the resource-based SHRM model, 'the value of the firm's strategic assets extends beyond their contribution to the production process' (p. 37).

Cappelli and Singh (1992) envision a coming 'marriage' between business strategy and HRM strategy based on the mutual recognition of the sustainable competitive advantage that skilled employees potentially create for the post-industrial organization. As Cappelli and Singh go on to discuss, this means that 'competitive advantage arises from firm-specific, valuable resources that are difficult to imitate' (p. 186). The strategic significance of HRM, Cappelli and Singh argue, is when HRM specialists demonstrate how, by developing valuable, non-transferable skills, HR impacts positively on long-term organizational performance. Similarly, Kamoche (1996) argues that when the two dimensions of 'human resource competencies' and the 'firm's core competencies' are aligned the 'full value of this synthesis is realisable' (p. 226). John Purcell (1995) also argues that the organization's HR assets can make the potential contribution of strategic HRM 'immense'.

## Limitations with the Resource-Based Model

How should we evaluate the resource-based model? As with the contingent matching approach, the resource-based approach to strategic HRM can be critiqued on both conceptual and empirical grounds. One problem is that the term itself, 'resource-based', appears to mean different things to different authors. Some competing terms include 'distinctive competence' (Selznick, 1957), 'core competencies' (Prahalad and Hamel, 1990), 'human resource competencies' and 'core competencies' (Kamoche, 1996), 'dynamic capabilities' (Teece *et al.*, 1997), and so on. The definitions range from narrow specific interpretations to very broad descriptions and are 'sometimes tautological; resources are defined as firm strengths, and firm strengths are then defined as strategic resources; capability is defined in terms of competence, and competence is then defined in terms of capability' (Nanda, 1996, p. 100).

The prescriptive validity of the resource-based approach has been questioned by some theorists. For some, this perspective to strategic HRM makes the mistake it accuses the matching model of making; 'it seems to be ascribing pre-eminence to the inside-out perspective of strategy' (Nanda, 1996, p. 103). The practitioner literature, however, warns against ignoring the strategic relevance of both external and internal factors and calls for a *'dual'* focus on market analysis and organizational capabilities. A second problem with the resource-based SHRM model stems from its implicit acceptance of a unitary perspective of the post-industrial workplace in which goals are shared and levels of trust are high. As is the case with the matching model, advocates of the resource-based SHRM perspective omit the dynamics of workplace trade unionism in the strategic equation. However, writers typically recognize the importance of workers' contribution to the labour process, knowledge and skills, synergy, proactive leadership, encouraging innovation and stimulating learning processes and, in contrast to the matching approach, it is a dynamic model of strategy.

What empirical support is there for the resource-based SHRM model? To date, the literature reports that empirical studies have lagged behind this model of strategic HRM. In relation to the question of empirical support, Nanda (1996) makes a pertinent observation: 'While the analysis has been sophisticated at macrotheoretic level, it stands relatively unsupported by microtheoretic foundations on the one side and empirical verifications on the other' (Nanda, 1996, p. 97). Further, Anglo-North American management have placed more emphasis on non-standard employment and reducing labour costs and de-emphasized explicitly or implicitly employment security which would suggest a limited acceptance of the resource-based approach as it applies to HRM (Boxall, 1996). In Canada, the general record of workplace training is 'dismal' (Bratton and Gold, 1999). Betcherman *et al.*'s (1994) survey study reported that 'the large majority of firms do not take a systematic, forward-looking approach to training; roughly 20 per cent appear to have a training budget and about 15 per cent have a formal training plan' (1994, p. 36).

## DIMENSIONS OF INTERNATIONAL STRATEGIC HRM

In addition to focusing on the validity of the matching and the resource-based SHRM models that have just been discussed, researchers have identified a number of important themes associated with the notion of strategic HRM that are briefly discussed in this section. These are: re-engineering organizations and work; leadership; workplace learning; trade unions, and the SHRM–performance link.

### Re-engineering and SHRM

The normative models of HRM emphasize the importance of organizational and job redesign. Much of the literature on the 'soft' HRM model is concerned with job design that would encourage the vertical and horizontal compression of tasks, worker autonomy and self-control or accountability. A new buzzword for the redesign of work organizations is 'business process re-engineering' (BPR). Hammer and Champy (1993) identify four features of BPR. First, the hierarchy of the corporation is 'flattened' as many middle-management positions give way to 'enabling' information technology and self-managed work teams. Second, work is redesigned into self-managed teams and managerial accountability is shifted to the 'front line'. Third, information technology is a 'critical enabler' that allows organizations to do work in 'radically' different ways. Fourth, senior management make an 'unwavering' commitment to radical change, including cultural change, set ambitious goals, and initiate the re-engineering process. This perspective to re-engineering process draws attention to 'strong' leadership and corporate culture, and the critical role of HRM.

### Leadership and SHRM

The concept of managerial leadership permeates and structures the theory and practice of work organizations and hence the way we understand SHRM. Most definitions of leadership reflect the assumption that leadership involves a process whereby an individual exerts influence upon others in an organizational context. More critical accounts of leadership tend to focus upon the hierarchical forms to which it gives rise, power relationships, and gender dominance (Smircich and Morgan, 1982; Townley, 1994). Within the leadership literature, there is a continuing debate over the alleged differences between a manager and a leader. Kotter (1990, 1996) proposed that managers develop *plans* whereas leaders create a *vision* and a strategy for achieving the vision.

The concept of leadership is a central building block of the resource-based SHRM model's concern with developing a 'strong' organizational culture and developing the organization's human intelligence and, moreover, generating employee commitment, flexibility, innovation and change (Bratton and Gold, 1999). The most explicit link between strategic HRM, workplace learning, and leadership is made by Senge (1990) when he writes that 'leaders are designers, stewards, and teachers' and that a learning organization will remain only a 'good idea, an intriguing but distant vision' until the leadership skills they require are more readily available (p. 340). Kotter (1996) also argues that the 'engine' that drives change is 'leadership, leadership, and still more leadership' (p. 32).

**Workplace Learning and SHRM**

In most formulations of SHRM, work-related learning has come to represent a key 'lever' that can help management achieve the substantive HRM goals of commitment, flexibility and quality. Many academics and corporate leaders have been attracted by the concept of the 'learning organization' (Cohen and Sproull, 1996), or 'workplace learning' (Spikes, 1995). Workplace learning is an inter-disciplinary body of knowledge and theoretical inquiry that draws upon adult learning and management theory. In practice, it is that part of the management process that attempts to facilitate work-related continuous learning at the *individual, group* and *organizational level*. Workplace learning occupies centre-stage in the resource-based SHRM model. Individual, team, and organizational learning can strengthen an organization's 'core competencies' and thus act as the engine for sustainable competitive advantage. From a managerial perspective, the pursuit of worker flexibility through workplace learning is discussed extensively by observers as a lever to sustainable competitive advantage: the ability to learn 'faster' than competitors (Dixon, 1992; Kochan and Dyer, 1995). There is also a growing body of work that has taken a more critical look at workplace learning (see Coopey, 1996).

**Trade Unions and SHRM**

In the literature the new HRM model is depicted as 'unitary'; it assumes that management and workers share common goals, and differences are treated and resolved rationally. In the prescriptive management literature, the argument is that the collectivist culture, with its 'them and us' attitude, sits uncomfortably with the HRM goal of high employee commitment and the individualization of the employment relationship, including individual contracts, communications, appraisal and rewards. Much of the critical literature also presents the new HRM model as inconsistent with traditional industrial relations and collective bargaining, albeit for very different reasons. Critics argue that HRM policies and practices are designed to obscure underlying sources of conflict inherent in employment relations (Godard, 1994). Others have argued that independent trade unions and variants of the HRM model can not only coexist, but are even necessary to its successful implementation and development. They argue that unions should become proactive or change 'champions' actively promoting the learning elements of the resource-based SHRM model. Such a union strategy would create a 'partnership' between management and organized labour which would result in a 'high-performance' workplace with mutual gains for both the organization and workers (Betcherman *et al.*, 1994, Verma, 1995; Guest, 1995).

## Organizational Performance and SHRM

Although most SHRM models provide no clear focus for any test of the HRM-performance link, the models tend to assume that an alignment between business strategy and HRM strategy will improve organizational performance and competitiveness. Do work organizations with a better 'fit' between HRM practices and business strategy have superior performance?

Some observers have suggested that there is little conclusive evidence of positive links between 'progressive' SHRM and higher productivity (see Purcell, 1989; Legge, 1995; Guest, 1997). There are still gaps in our knowledge, but North American scholars have recently provided important information on these empirical questions. Ichniowski, Kochan, Levine, Olson and Strauss (1996) review the findings from a body of US research, and Betcherman *et al.* (1994) provides evidence on the HRM-organizational performance relationship using Canadian data. Both Ichniowski *et al.* and Betcherman *et al.* argue that the research evidence suggests that innovative HRM practices can increase organization performance.

The work by Ichniowski *et al.* reviews a diverse body of research on the HRM–firm performance link. Longitudinal case studies in a California-based auto assembly plant and a US paper mill document the reconfiguration of traditional work structures to the 'team concept' and subsequent improvements in productivity and quality performance. A cross-sectional comparative case study of two apparel garment factories found that 'team-orientated' work structures produced a 30 per cent advantage in overall production costs over a traditional work structure. The results do need interpreting with some caution. The performance measures differ across studies and so are not comparable. Further, access to performance data may suggest that the more successful firms are over-represented (Ichniowski *et al.*, 1996). The findings from four intra-industry studies – steel making, automobile assembly, apparel manufacture and metalworking – show that different work configurations and worker empowerment arrangements associated with the new HRM model have superior output and quality performances. Of particular interest is the finding that *integrated* HRM innovations have a greater effect than individual HRM practices.

Ichniowski *et al.*'s main conclusion is that 'There are no one or two "magic bullets" that are *the* work practices that will stimulate worker and business performance. Work teams or quality circles alone are not enough. Rather, *whole systems* [my emphasis] need to be changed' (1996, p. 322). The evidence suggests an apparent paradox. If the pursuit of core HR practices leads to improved organizational performance, why aren't such practices more widely used? This may result from the long-term investment costs associated with the resource-based approach to strategic HRM and the pressure on individual managers to achieve short-term financial results.


## TOWARDS AN ASIAN MODEL OF SHRM?

In this final section we address aspects of the international scene to help us put the SHRM discourse in an Asian context. In doing so, we make a distinction between international HRM and comparative HRM. The subject-matter of the former revolves around the issues and problems associated with the globalization of capitalism. Comparative HRM, on the other hand, focuses on providing insights into the nature of, and reasons for, differences in HRM practices across national boundaries. This section addresses a number of questions which will further develop our knowledge and understanding of SHRM. Can the Anglo-American SHRM model be transplanted to Asian organizations? To what extent are HRM practices culturally determined? In terms of national models, is it feasible to speak of a distinctively 'Asian HRM model'?

So far, the majority of *international HRM* research has focused on issues associated with the cross-national transfer of people – for instance, how to select and manage expatriate managers in international job assignments (e.g., Tung, 1988; Shenkar, 1995). It has been argued elsewhere that much of this work is descriptive, managerialist and lacks analytical rigour (e.g., Kochan *et al.*, 1992).

International HRM tends to emphasize the subordination of national culture and national employment practices to corporate culture and HRM practices (Boxall, 1995). The issue of transplanting Western HRM practices and values into culturally diverse environments needs to be critically researched. Early twenty-first-century capitalism, when developing international business strategy, faces the perennial difficulty of organizing the employment relationship to reduce the indeterminacy resulting from the unspecified nature of the employment contract (Townley, 1994). If we adopt this approach to international HRM, the role of knowledge to render men and women in the workplace 'governable' is further complicated in culturally diverse environments. For example, it behoves researchers to examine whether managers and workers in China, Hong Kong, India, Pakistan, Singapore, and Thailand and elsewhere in Asia will accept the underlying ideology and embrace the SHRM paradigm.

In terms of critical research the field of study referred to as *comparative HRM* is also relatively underdeveloped. Of considerable interest to academics and practitioners is the question of the extent to which HRM practices that work effectively in one country and culture can be transplanted to others. Drawing upon Bean's (1985) work on comparative HRM it is defined here as a systematic method of investigation relating to two or more countries, which has analytic rather than descriptive implications. On this basis, comparative HRM should involve activities that seek to explain patterns and variations encountered in cross-national HRM rather than a simple description of HRM institutions and practices in selected countries (Shalev, 1981). Comparative studies can lead to a greater understanding of the factors and processes which determine HRM phenomena. The common assumption found in many North American management textbooks on HRM is that 'best' practice has universal application. The assumption of universal deployment is untenable since the HRM phenomena reflect different cultural milieu (Boxall, 1995; Martin and Nakayama, 1999). Comparative HRM studies can provide the basis for a 'revolution' in management philosophy and practices by offering 'lessons' from off-shore experience. Comparative HRM studies can also challenge the rhetoric of the more prescriptive international HRM literature. It can promote wider understanding of, and foster new insights into HRM.

Using comparative analysis, Moore and Devereaux Jennings (1995) examine the HRM paradigm from a North American perspective. Drawing upon the data from twelve countries on the Pacific Rim, they illustrate diversity in the application of IIR practices both *between* countries located in Asia and diversity between work organizations *within* nation states. For example, in the important area of union–management relations, Japanese industrial relations are characterized as 'cooperative' (Morishima, 1995). This is in contrast to Taiwan where the government has a 'deep-seated fear of labour unrest' and consequently trade unions are subject to repressive government control (Farh, 1995, p. 265). There are examples of considerable diversity in individual nation states. In Singapore, for instance, it is reported that there is diversity in the 'microcosm' due to the influence of local traditions (Chinese, Indian, and Malay), the presence of a large number of multinational companies from different countries – Asian, North American and European – and the mix of small family businesses and large public organizations (Moore and Devereaux Jennings, 1995, p. 262). Similarly, in Thailand, there appears to be considerable diversity in HR practices in publicly owned organizations and traditional Thai family enterprises.

Inherent in controversies around the notion of an 'Asian model' are questions of the limitations and value of cross-national generalizations in human resource management (Hyman, 1994). The diversity in the application of HR practices in countries located in the Asian Region raises, amongst other issues, the importance of intercultural communications and cultural idiosyncrasies. International SHRM (ISHRM) presents challenges to language and non-verbal communication. Metaphors are commonly used by Western managers but many may be incompatible in different contexts (Martin and Nakayama, 1999). While language communicates explicit information, non-verbal communication conveys relational signals – how we feel about a person – and also communicate status and power. For example, eye contact communicates meanings about respect and status, and patterns of eye contact vary between cultures.

Communication and culture are both interrelated and reciprocal (Martin and Nakayama, 1999), and this dialectic process should help HR practitioners and researchers understand why researching the deployment of HR practices is problematic. According to Moore and Devereaux Jennings (1995), core sets of HR practices which appeared similar on the 'surface' when introduced into Asian organizations, were different *below* the surface. Building on Tyson and Brewster's (1991) hypothesis, we may have to acknowledge the existence of discrete 'HRM models' both between and within nations, contingent on distinct economic and cultural factors. The diversity in the application of HR practices raises at least two other questions: What elements of the SHRM paradigm are truly imitable and which are not? Does the non-deployment of SHRM core practices undermine competitive advantage? The motivation to *attempt* to transplant core HR practices is partly dependent upon the answer to the second question. Evidence has been provided that innovative HRM practices do increase organizational performance. The argument that the revolutionary change in practices – from pragmatic, reactive employment management to the sophisticated strategic resource-based model – cannot be brought about by managers alone (see Kochan *et al.*, 1986; Sisson, 1994; Storey, 1995). It tends to favour those Asian countries that have made a strategic choice to adopt an HRM system that includes other stakeholders. Such an alliance or 'partnership' is more conducive to cooperation and the development of strategic intangibles: human intelligence and learning. It is easier to formulate questions than answers, and this section has taken the easier route rather than the more difficult. Yet there is value in asking questions. Questions can stimulate reflection and increase our understanding of the ISHRM paradigm. My object has been to do both.


## SUMMARY

This chapter has examined different level of strategic management. In multi-divisional corporations, strategy formulation takes place at three levels – corporate, business, and functional – to form a hierarchy of strategic decision-making. We draw attention to the more critical literature that recognizes that the strategic HRM options at any given time are partially constrained by the outcomes of corporate and business decisions, the current distribution of power within the organization, and the ideological values of the key decision-makers. Whether senior managers adopt the 'matching' or the 'resource-based' model of SHRM will be contingent upon the corporate and business strategies, as well as by varying degrees of pressure and constraints from environmental forces. Finally, we differentiated between international HRM and comparative HRM. Using an interdisciplinary approach, it was posited that to bring about a 'revolution' in HR management, stakeholders have to recognize, *inter alia*, the importance of intercultural communications and Asian cultural idiosyncrasies.


**References**

Aktouf, O. (1996) *Traditional Management and Beyond* (Montreal: Morin).
Amit, R. and Shoemaker, P. J. H. (1993) 'Strategic Assets and Organizational Rent', *Strategic Management Journal*, 14, pp. 33–46.
Bamberger, P and Phillips, B. (1991) 'Organizational Environment and Business Strategy: Parallel versus Conflicting Influences on Human Resource Strategy in the Pharmaceutical Industry', *Human Resource Management*, 30, pp. 153–82.
Barney, J. B. (1991) 'Firm Resources and Sustained Competitive Advantage', *Journal of Management*, 17(1), pp. 99–120.

Bean, R. (1985) *Comparative Industrial Relations* (London: Croom Helm).

Beer, M., and Eisenstat, R., (1996) 'Developing an Organization Capable of Implementing Strategy and Learning', *Human Relations*, 49(5), pp. 597–619.

Beer, M., Spector, B., Lawrence, P. R., Quin Mills, D. and Walton, R. E. (1984) *Managing Human Assets* (New York: Fress Press).

Betcherman, G., McMullen, K., Leckie, N. and Caron, C. (1994) *The Canadian Workplace in Transition* (Queen's University, Kingston, Ontario: IRC Press).

Boxall, P. F. (1992) 'Strategic Human Resource Management: Beginnings of a New Theoretical Sophistication?', *Human Resource Management Journal*, 2(3).

Boxall, P. (1995) 'Building The Theory of Comparative HRM', *Human Resource Management Journal*, 5(5), pp. 5–17.

Boxall, P. (1996) 'The Strategic HRM Debate and the Resource-Based View of the Firm', *Human Resource Management Journal*, 6(3), pp. 59–75.

Bratton, J. and Gold, J. (1999) *Human Resource Management: Theory and Practice* (London: Macmillan).

Cappelli, P. and Singh, H. (1992) 'Integrating Strategic Human Resources and Strategic Management', in D. Lewin, O. S. Mitchell and P. Sherer (eds), *Research Frontiers in Industrial Relations and Human Resources* (Madison, WI: Industrial Relations Research Association) pp. 165–92.

Champy, J. (1996) *Reengineering Management: The Mandate for New Leadership* (New York: Harper Collins).

Chandler, A. (1962) *Strategy and Structure* (Cambridge, Mass: MIT Press).

Child, J. (1972) 'Organizational Structure, Environment and Performance: The Role of Strategic Choice', *Sociology*, 6(1).

Cohen, M. D. and Sproull, L. S. (eds) (1996) *Organizational Learning* (Thousand Oaks: Sage).

Coopey, J. (1996) 'Crucial Gaps in "the Learning Organisation": Power, Politics and Ideology', in K. Starkey (ed.), *How Organisations Learn* (London: International Thomson Publishing).

Daft, R. (1998) *Organization Theory and Design*, 6th edn (Cincinnati, Ohio: South-Western).

Devanna, M. A., Fombrun, C. and Tichy, N. (1984) 'A Framework for Strategic Human Resource Management', in Fombrun C. *et al.*, *Strategic Human Resource Management* (Chichester: John Wiley).

Dicken, P. (1992) *Global Shifts*, 2nd edn (London: Guilford Press).

Dixon, N. (1992) 'Organizational Learning: A Review of the Literature with Implications for HRD Professionals', *Human Resource Development Quarterly*, 3(1), pp. 29–49.

Downie, B. and Coates, M. (1994) *Traditional and New Approaches to Human resource Management* (Kingston, Ontario: IRC Press).

Dyer, P. (1984) 'Studying Human Resource Strategy: An Approach and an Agenda', *Industrial Relations*, 23(2).

Farh, Jiing-Lih (1995) 'Human Resource Management in Taiwan, The Republic of China', in L. Moore and P. Devereaux Jennings (eds), *Human Resource Management on the Pacific Rim* (New York: Walter de Gruyter) pp. 265–94.

Fombrun, C., Tichy, N. M. and Devanna, M. A. (eds) (1984) *Strategic Human Resource Management* (New York: Wiley).

Godard, J. (1994) *Industrial relations: The Economy and Society* (Toronto: McGraw-Hill Ryerson).

Guest, D. E. (1987) 'Human Resource Management and Industrial Relations', *Journal of Management Studies*, 24(5).

Guest, D. (1995) 'Human Resource Management, Trade Unions and Industrial Relations', in John Storey (ed.), *Human Resource Management: A Critical Text* (London: Routledge).

Guest, D. (1997) 'Human Resource Management and Performance: A Review and Research Agenda', *International Journal of Human Resource Management*, 8(3), pp. 263–76.

Hammer, M. and Champy, J. (1993) *Reengineering the Corporation* (London: Nicholas Bealey).

Harris, P. and Moran, R. (1991) *Managing Cultural Differences* (London: Gulf Publications).

Hill, C. and Jones, G. (1998) *Strategic Management Theory* (Toronto: Houghton-Mifflin).

Hyman, R. (1994) 'Industrial Relations in Western Europe: An Era of Ambiguity?', *Industrial Relations Journal*, 33(1), pp. 1–24.

Ichniowski, C., Kochan, T., Levine, D., Olson, C. and Strauss, G. (1996) 'What Works at Work: Overview and Assessment', *Industrial Relations*, 35(3), pp. 299–333.

Johnson, G. (1987) *Strategic Change and the Management Process* (Oxford: Blackwell).

Kamoche, K. (1996) 'Strategic Human Resource Management within Resource-Capability View of the Firm', *Journal of Management Studies*, 33(2), pp. 213–33.

Kochan, T., Katz, H. C. and McKersie, R. B. (1986) *The Transformation of American Industrial Relations* (New York: Basic Books).

Kochan, T., Dyer, L. and Batt, R. (1992) 'International Human Resource Studies: A Framework for Future Research', in D. Lewin *et al.* (eds), *Research Frontiers in Industrial Relations and Human Resources* (Madison, WI: Industrial Relations Research Association) pp. 309–38.

Kochan, T. and Dyer, L. (1995) 'HRM: An American View', in John Storey (ed.), *Human Resource Management: A Critical Text* (London: Routledge).

Kotter, J. (1990) *A Force for Change* (New York: Free Press).

Kotter, J. (1996) *Leading Change* (Boston, Mass.: Harvard Business).

Kydd, B. and Oppenheim, L. (1990) 'Using Human Resource Management to Enhance Competitiveness: Lessons from Four Excellent Companies', *Human Resource Management Journal*, 29(2), pp. 145–66.

Legge, K. (1995) *Human Resource Management: Rhetorics and Realities* (Basingstoke: Macmillan).

Martin, J. and Nakayama, T. (1999) *International Communications in Contexts* (Mountain View, CA: Mayfield Publishing).

McLoughlin, I. and Clark, J. (1988) *Technological Change at Work* (Milton Keyes: Open University Press).

Miles, R. and Snow, C. (1984) 'Designing Strategic Human Resources Systems', *Organizational Dynamics*, Summer, pp. 36–52.

Miller, P. (1987) 'Strategic Industrial Relations and Human Resource Management – Distinction, Definition and Recognition', *Journal of Management Studies*, 24(4).

Mintzberg, H. (1994) *The Rise and Fall of Strategic Planning* (Hemel Hempstead: Prentice Hall).

Moore, L. F. and Devereaux Jennings, P. (eds) (1995) *Human Resource Management on the Pacific Rim: Institutions, Practices and Attitudes* (New York: Walter de Gruyter).

Morishima, M. (1995) 'The Japanese Human Resource Management System: A Learning Bureaucracy', in L. Moore and P. Devereaux Jennings (eds), *Human Resource Management on the Pacific Rim* (New York: Walter de Gruyter) pp. 119–50.

Nahavandi, A. and Malekzadeh, A. R. (1993) 'Leader Style in Strategy and Organizational Performance: An Integrative Framework', *Journal of Management Studies*, 30(3), pp. 405–25.

Nanda, A. (1996) 'Resources, Capabilities and Competencies', in B. Moingeon and A. Edmondson (eds), *Organisational Learning and Competitive Advantage* (London: Sage).

Parker, B. (1999) 'Evolution and Revolution: From International Business to Globalisation', in S. Clegg *et al.*, *Managing Organisations* (London: Sage).

Peck, S. (1994) 'Exploring the Link between Organizational Strategy and the Employment Relationship: The Role of Human Resource Policies', *Journal of Management Studies*, 31(5), pp. 715–36.

Penrose, E. (1959) *The Theory of the Growth of the Firm* (Oxford: Blackwell).

Porter, M. (1980) *Competitive Strategy* (New York: Free Press).

Porter, M. (1985) *Competitive Advantage: Creating and Sustaining Superior Performance* (New York: Free Press).

Prahalad, C. K. and Hamel, G. (1990) 'The Core Competencies of the Organization', *Harvard Business Review*, 68 (May–June), pp. 79–91.

Purcell, J. (1989) 'The Impact of Corporate Strategy on Human Resource Management', in J. Storey (ed.), *New Perspectives on Human Resource Management* (London: Routledge).

Purcell, J. (1995) 'Corporate Strategy and its Link with Human Resource Management Strategy', in John Storey (ed.), *Human Resource Management: A Critical Text* (London: Routledge).

Purcell, J. and Ahlstrand, B. (1994) *Human Resource Management in the Multi-Divisional Company* (Oxford: Oxford University Press).

Robertson, R. (1995) 'Glocalisation: Time–Space and Homogeneity–Hetrogeneity', in M. Featherstone, S. Lash and R. Robertson (eds), *Global Modernities* (London: Sage) pp. 25–44.

Schuler, R. S. (1989) 'Strategic Human Resource Management and Industrial Relations', *Human Relations*, 42(2).

Schuler, R. and Jackson, S. (1987) 'Linking Competitive Strategies and Human Resource Management Practices', *Academy of Management Executive*, 1(3), pp. 209–13.

Selznick, P. (1957) *Leadership and Administration* (New York: Harper & Row).

Senge, P. (1990) *The Fifth Discipline* (New York: Doubleday).

Shalev, M. (1981) 'Limits of and Alternatives to Multiple Regression in Macro-Comparative Research', paper prepared for presentation at the Second Conference on *The Welfare State at the Crossroads* (Stockholm, June 12–14).

Shenkar, O. (1995) *Global Perspectives on Human Resource Management* (New York: Prentice Hall).

Sisson, K. (ed.) (1994) *Personnel Management*, 2nd edn (Oxford: Blackwell).

Smircich, L. and Morgan, G. (1982) 'Leadership: The Management of Meaning', *Journal of Applied Behavioral Science*, 18(3), pp. 257–73.

Spikes, W. F. (ed.) (1995) 'Workplace Learning', *New Directions for Adult and Continuing Education*, 68 (Winter) (San Francisco: Jossey-Bass).

Storey, J. (1995) 'Human Resource Management: Still Marching On or Marching Out?', in J. Storey (ed.), *Human Resource Management: A Critical Text* (London: Routledge).

Storey, J., Cressey, P., Morris, T. and Wilkinson, A. (1997) 'Changing Employment Practices in UK Banking: Case Studies', *Personnel Review*, 26(1), pp. 24–42.

Teece, D. J., Pisano, G. and Shuen, A. (1997) 'Dynamic capabilities and strategic management', in N. J. Foss (ed.), *Resources, Firms* (Oxford: Oxford University Press) pp. 268–85.

Thurley, K. and Wood, S. (1983) 'Business Strategy and Industrial Relations Strategy' in K. Thurley and S. Wood (eds), *Industrial Relations and Management Strategy* (Cambridge: CUP).

Tichy, N. and Devanna, M. (1986) *The Transformational Leader* (New York: John Wiley).

Townley, B. (1994) *Reframing Human Resource Management* (London: Sage).

Tung, R. L. (1988) *The New Expatriates: Managing Human Resources Abroad* (Cambridge, MA: Ballinger).

Tyson, S. and Brewster, C. (1991) 'Comparative Studies and the Development of Human Resource Management', in C. Brewster and S. Tyson (eds), *International Comparisons in Human Resource Management* (London: Pitman) pp. 257–9.

Verma, A. (1995) 'Employee Involvement in the Workplace', in M. Gunderson and A. Ponak (eds), *Union–Management Relations in Canada*, 3rd edn (Don Mills, Ontario: Addison-Wesley) pp. 281–308.

Watson, T. (1999) 'Human Resource Strategies: Choices, Chance and Circumstances', in J. Leopold *et al.* (eds), *Strategic Human Resourcing* (London: Pitman).

Wheelen, T. and Hunger, J. (1995) *Strategic Management and Business Policy*, 5th edn (New York: Addison-Wesley).

Whittington, R. (1993) *What is Strategy and Does it Matter?* (London: Routledge).

# 3 Managing Employee Competence in Global Organizations

Ronald L. Jacobs

## INTRODUCTION

Recent economic uncertainties in Asia have prompted many managers to reflect on their future actions regarding employee competence (Diamond and Plattner, 1998). The prevailing sense is that future global competitiveness depends in large part on organizations having employees with higher levels of competence (Friedman, 1999). While there seems general agreement about the importance of employee competence, there seems less certainty about how employee competence might actually be understood and managed. That is, if a more skilled workforce is required in organizations, then how do managers specifically address this issue in practical terms? Thus, understanding employee competence has taken on greater importance, especially in regions undergoing dynamic economic change.

This chapter is divided into four sections. The first section discusses human competence and its relationship with organizational performance. The second section describes a hierarchy that shows five levels of employee competence. The third section discusses how to acquire different levels of competence and the need to match competence level with organizational requirements. Finally, the chapter presents implications related to managing employee competence.

## HUMAN COMPETENCE AND PERFORMANCE

Managers can do many things to improve the performance of their organizations. They can bring in advanced technologies, streamline production and service delivery processes, introduce new products and services, and change the work design. But when all is said and done, the one managerial action that ultimately determines organizational success is ensuring that employees can perform what is expected of them. That the most skilful employees make a disproportionate impact on organizational performance has become an accepted fact. Indeed, without adequate levels of human competence, the ability of organizations to undertake change is severely limited. Jacobs and Jones (1995) have observed this relationship by stating, 'Organizations must transform themselves to become more competitive, and the know-how of employees has become more critical for ensuring the success of the transformation process' (p. 5). Herling and Provo (2000) note that business success increasingly hinges on an organization's ability to use its employees' expertise as a factor in the shaping of its business strategy. Increasingly, organizations are beginning to realize how their market value relies on the knowledge and skills of their employees and that caring about their human competency base and how it is developed is starting to make strategic sense to them.

In practice, human competence has become an illusive concept. No one definition seems able to define the concept in all its varied and rich aspects. Gilbert (1978) suggested that human competence is about an individual's potential to use a set of knowledge and skills in a specific situation. In general, human competence has come to have four defining principles (Bereiter and Scardamalia, 1993). First, if human competence is concerned with abilities, then it must have a particular context. Thus, human competence always describes a range of abilities relative to a particular set of work or tasks. Second, human competence does not necessarily describe the learning process by which individuals attained the abilities. Rather, it emphasizes what performance outcomes individuals can potentially achieve as a result of having the abilities. Third, measuring human competence should be done through the use of clearly defined standards, such that judgments about an individual's level of competence should be based on comparisons with an objective standard. Finally, human competence is a measure of what someone can do at one particular point in time. Past performance might predict future performance in some instances, but not in organizational situations in which competence requirements undergo constant change.

Perhaps some of the confusion surrounding human competence stems from the range of organizational situations in which individuals are asked to demonstrate their abilities. Logically, individuals differ from each other in their use of technical knowledge, in solving certain types of production or human resource problems, or in communicating ideas to others. Whatever the situation, human competence describes the individual's potential to use what they know and can do to meet the demands of that narrowly defined situation. Thus, human competence has a value-free connotation.

As stated, there is a fundamental relationship between human competence and organizational performance. In general, the more competence available among employees should relate to higher levels of organizational perform-ance. But for that relationship to hold true, the capabilities of the employees must match the requirements of the organization. Achieving this match at all times has become more and more difficult for at least two related reasons. First, as more and more organizations have moved to a process orientation, the context of work has undergone fundamental change. For instance, many organizations at the Hshinchu Industrial Park, in Taiwan, have moved to an organizational structure driven by cross-functional workflows rather than a traditional functional management structure. As such, several authors have suggested the end of the job as the means for conceptualizing an individual's contribution in organizations (Davenport, 1993; Rummler and Brache, 1995).

That is, jobs were traditionally understood to have distinct boundaries based on their exclusivity from other jobs. Job boundaries were delineated by specifying sets of unique job duties and tasks that differentiated the expect-ations of one employee from another. Thus, individuals in one job were expected to perform their assigned tasks relatively independent from individ-uals in the second job. Such boundaries existed even when in fact both employees might have contributed products or services to the same work process in a supplier–customer relationship.

Although rigid job boundaries have served to protect workers' rights in the past, such approaches have also come to hinder efforts to view organizations as dynamic systems with interconnecting improved organizational perform-ance, a notion that does not have to contradict the protection of workers' rights. Indeed, most organization improvement efforts have consisted of giving employees, especially frontline employees, greater autonomy and flex-ibility in doing their work. However, establishing boundaries between jobs has not been entirely eliminated in many organizations, but they have been recon-figured based on another set of considerations. Job boundaries now seem based more on the expectation of solving different sets of problems and making decisions related to the same set of tasks. In this sense, more and more job boundaries are being made based on the natural classes of problems or decisions that emerge related to performing tasks, rather than artificial differences in job role.

Second, there is the general understanding that regardless of how jobs are conceived, job content has continued to undergo rapid change. Few jobs in the new economy are expected to remain the same for any length of time. For instance, managers and frontline employees among hard-disk manufacturers in the northern part of Singapore, such as Seagate, realize that certain production tasks change on a weekly basis, if not faster. Indeed, if the tasks do not change on a continual basis, then those organizations might suffer because of the lack of innovation that those changes would bring about, which would

*Table* 3.1   Hierarchy of employee development

| Level | Description |
| --- | --- |
| Novice | Literally, one who is new to a particular work situation. There is often some, but minimal exposure to the task beforehand. As a result, the individual lacks the knowledge and skills necessary to meet the requirements set to adequately perform the task. |
| Specialist | One who can reliably perform specific tasks unsupervised. But the range of tasks is limited to the most routine ones. Often it is necessary to coach individuals at this level to help them use the most appropriate behaviours. |
| Experienced specialist | One who can perform specific tasks and who has performed those tasks many times. As a result, the individual can perform the tasks with ease and skill. It is possible to remain at this level for an extended period of time. |
| Expert | One who has the knowledge and experience to meet and often exceed the requirements of the task. The individual is distinguished and highly regarded by peers because of his or her consummate skills, or expertise. The individual can use this ability to deal with routine and non-routine cases, with an economy of effort. |
| Master | One who is regarded as 'the' expert among experts, or expert compared to others employees. He or she is among the elite group whose judgments are looked upon to set the standard and ideals for others. |

*Source*:   Jacobs (1997).

affect their productivity in the long run. In such a situation, managers realize that the notion of employee competence becomes a dynamic notion.

## HIERARCHY OF EMPLOYEE COMPETENCE

As shown in Table 3.1, Jacobs (1997) described five levels of employee competence that an individual might possess relative to a specific set of work: (1) novice, (2) specialist, (3) experienced specialist, (4) expert, and (5) master. Each level represents an increasingly more complex amount of knowledge and skill than the level preceding it. In this sense, employee competence is a relatively narrow concept, primarily defined by its referent back to the work. Indeed, it is possible that an individual might be considered as performing both at the novice and expert levels at the same time, but on different tasks within their area of responsibility.

- *Novice.* Novice describes a level of competence in which employees are unable to perform a certain task simply because they lack the necessary knowledge, skills, and experience to do so. The novice level is relatively easy to identify in organizations, as this level represents individuals who have been newly hired, recently transferred, or have been recently promoted into new positions.
- *Specialist.* In some respects, the specialist level is perhaps the most frustrating level of employee competence for managers. On the one hand, by design, employees with this level of competence in a task can make important contributions to their organizations because they can reliably perform routine tasks. That is, they can perform known tasks such as complex quality assurance on finished hard disks. On the other hand, because they have limited information, they lack the ability to be effective in doing anything much beyond the routine tasks. They cannot deviate from what they have been specifically trained to perform.
- *Experienced Specialist.* The experienced specialist level describes when an individual has had many opportunities to perform and practise the information learned through training and on-the-job experiences. As a result, the individual can now do the task with some level of skill. Individuals performing a task at the experienced specialist level are often mistaken for having greater abilities than they actually possess, simply because of the ease in which they perform tasks. Even so, the experienced specialist level lacks the combination of knowledge and experience to be truly considered at the expert level.
- *Expert.* A few individuals at the experienced specialist level have the opportunity to move to the expert level of competence on a task. For a variety of reasons, not all individuals can achieve this level of development. Being an expert means more than being good at the task alone. It has been shown that the expert level is reserved for individuals who are able to respond to unique variations of the task, drawing from deep knowledge and experience. Indeed, experts have been known to derive more creative solutions to common problem situations (Robinson and Stern, 1997).

By definition, being an expert means that these individuals truly have distinct ways of organizing and using information that differ substantially from others around them (Chi, Glaser and Farr, 1988). Among other behaviours, experts have unique ways of viewing and solving problems. Experts have a propensity to engage in progressive problem solving – a technique by which they create problem situations in which none existed beforehand from which hypotheses can be generated and confirmed. In this way, at the expert level, individuals intentionally impose theory on to problems, with the intent of enriching their own understanding of the situation.

- *Master.* Individuals at the master level represent a particularly elite group of experts. Their actions and judgments often set the standards for all others beneath them. Master employees are the individuals who are regarded as 'the' expert among experts. They are the role models from which others can learn and judge the adequacy of their own knowledge and skills. Others turn to the master employee when an interpretation of information is required. Master employees are often asked to be the trainers of others, but their high level of knowledge and experience may be unnecessary for many training situations.

## PLANNING FOR EMPLOYEE COMPETENCE NEEDS

Clearly, managers must assume the responsibility of addressing employee competence issues in their organizations. Table 3.2 shows a list of human resource development solutions that organizations use to achieve certain levels of employee competence. Formal training programmes, such as class-room training or structured on-the-job training, provide the most reliable means to enable employees achieve to the specialist level of competence. Of course, individuals can learn to this level on their own, but the outcomes of using this approach are not always predictable (Jacobs, Jones and Neil, 1992). To move from the specialist to the experienced specialist level, individuals must be given a range of experiences to use what was learned during the training. Formal and informal training opportunities, such as team meetings, mentoring, and coaching, would be appropriate means to help move specialists to the experienced specialist level of competence (Watkins and Marsick, 1994). These approaches must be used in conjunction with grounded experience for them to have any meaning.

As stated, individuals who can perform a task at the expert level possess characteristics distinct from experienced specialists, and the confusion between the two levels often has detrimental affects on organizational per-formance. How to become an expert is subject to some conjecture, some of which depends more on the individuals than the task. However, there seems general agreement that achieving to the expert level of competence requires deep knowledge about the task and a breadth of experiences in applying that knowledge. Thus, external educational opportunities and opportunities to mentor others might be used to begin this process more reliably. Unfortu-nately, in practice, organizations are usually less able to develop experts on a reliable basis than the previous levels of competence. However, when an

*Table* 3.2   Competence levels and human resource development solutions

| *Levels* | *HRD solutions* |
| --- | --- |
| Specialist | • On-site classroom training programmes<br>• Structured on-the-job training programmes<br>• Off-site training courses<br>• Structured coaching sessions<br>• Guided practice sessions |
| Experience specialist | • On-site classroom training programmes<br>• Structured on-the-job training<br>• Focused experience in doing task<br>• Impromptu coaching sessions<br>• Team meetings |
| Expert | • Wide range of experience in doing task<br>• Off-site training and educational programmes<br>• Disposition to ask questions/seek solutions<br>• Impromptu reflection-in-action discussions<br>• Team meetings |
| Master | • Disposition to help others learn task<br>• Off-site training and educational programmes<br>• Impromptu reflection-in-action discussions |

individual demonstrates a potential to achieve at the expert level, managers need to recognize and nurture those abilities over time.

In planning for the competence needs of employees, the question often arises whether organizations should attempt to develop all employees to the expert level of competence on all tasks? After all, it seems axiomatic that if employee competence is related to organizational performance, then even higher levels of competence would logically lead to higher levels of performance. From a practical perspective, attempting to develop all employees to the expert level might not be appropriate. For one thing, the organization might not require that individuals perform a given task to that level. Doing a task to the specialist or experienced levels might in fact be good enough in order to meet the expectations of the organization and customers. For another reason, some jobs inherently have high turnover rates or provide employees a pass-through to other jobs in the organization. As a result, there would seem little financial justification for developing individuals to higher levels of competence beyond what is immediately required.

In addition, developing individuals for levels of competence beyond what is required might in fact have a detrimental effect on expected performance.

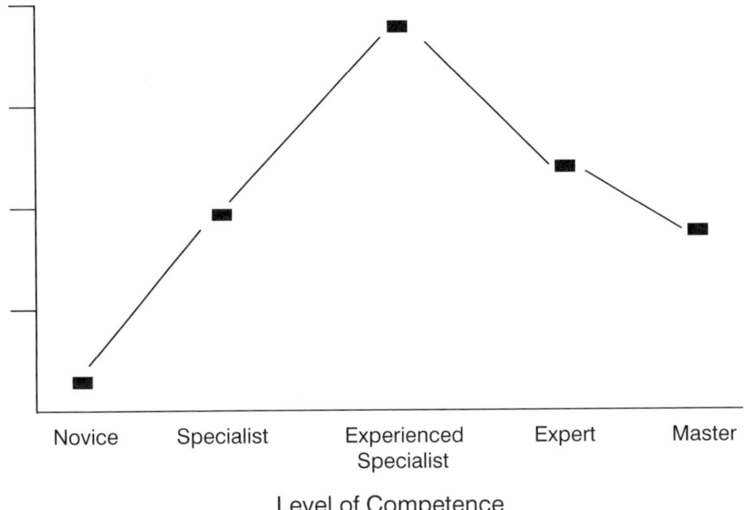


*Figure* 3.1    Relationship between performance and level of competence

For example, Figure 3.1 shows an example relationship between levels of employee competence and job performance outcomes. The figure shows that at the experienced specialist level of competence, job performance begins to diminish, not to continue to rise. Indeed, common sense suggests that the relationship between employee competence and organizational performance might be viewed as being more curvilinear than linear. For example, if achieving product quality outcomes depends on having employees follow specific steps of a procedural task without any deviation, then the specialist level of competence might be sufficient for the needs of the organization. However, whether the same employees would be equipped at this level to respond to unique situations that invariably arise is open to question.

## IMPLICATIONS FOR MANAGEMENT

At least implications are of importance to managers related to addressing employee competence issues. The first implication is that there is a need to analyse and record the components of critical work tasks. Jacobs and Jones (1995) present a list of task categories to help guide in the documentation process. The tasks should also be analysed with particular emphasis on the level of competence that is required to meet organizational requirements

(Swanson, 1996). In practice, this often requires greater attention, to the extent that problem-solving and decision-making occurs on known and unknown situations.

In general, the analysis of known problems, such as when the causes and corrective actions are commonly known, can be accomplished at the specialist level through formal training programmes and some experience. However, when problems occur in which the specific causes and actions are unknown, then such situations logically require a higher level of competence. Knowing the extent to which tasks can be expected to involve either known or unknown problems is a management requirement.

The second implication is that employee competence has increased from being exclusively a management responsibility to becoming a major force for the shaping of business strategy. As such, all levels of management should consider becoming more proactive in addressing employee competence issues. Resources in time, materials, and people need to be provided to ensure that formal and informal learning opportunities occur, appropriate for the levels of required competence. Unfortunately, this implication sounds easier to agree to in principle than to implement on an actual basis.

The challenge for many managers often comes down to the competing need of meeting immediate production and service delivery requirements on the one hand and taking time to invest in the organization's future capabilities on the other. If addressing employee competence does not exist as a critical business process, then the future viability of the organization might be called into question in the first place, regardless of the products or services provided.

The final implication suggests the need to view employee development in ways that benefit both the organization and frontline individuals. Jones and Jacobs (1997) suggest that there is an increasing need to focus greater attention on the development needs of those closest to the production of the product or delivery of the service. Frontline employees describe a unique category of employee that transcends other commonly used categories, such as production, managerial, or professional. Indeed, frontline employees include a range of job levels such as service employees, salespersons, engineers, health-care specialists. Frontline employees share the common feature of making things or providing services, as opposed to supervising other people, as the major part of their work.

Multi-skilling has emerged as one of the major challenges related to frontline employees. That is, how to ensure that these individuals can perform a wider range of tasks than what was considered typical in the past. Jones and Jacobs (1997) propose a five-stage employee develop-ment system specifically designed for frontline employees entering new work areas: (1) Define the process; (2) Conduct work-area orientation; (3) Prepare development plan; (4) Identify developmental approaches; and (5) Manage performance. The critical aspect for success is the recognition that both organization goals and individual goals need to be achieved as result of using the system. Employee development exists as a major way for employee competence issues to be addressed systematically, across all levels of employees.

## CONCLUSION

This chapter makes the point that managers should pay close attention to the employee competence needs of their organizations. However, to be effective, managers should understand that employee competence varies by its level of complexity, and that not all tasks require the same level to meet organizational requirements. Thus, managers should plan for the competence needs of their employees, similar to addressing other organizational issues. How to address the employee competence requirements of organizations remains a challenge across all organizations in the global economy, if not more so for organizations in highly competitive environments such as Asia.

## References

Bereiter, C. and Scardamalia, M. (1993) *Surpassing Ourselves: An Inquiry into the Nature and Implications of Expertise* (Chicago: Open Court).
Chi, M. T., Glaser, R. and Farr, M. J. (1988) *The Nature of Expertise* (Hillsdale, NJ: Lawrence Erlbaum Associates).
Davenport. T. (1993) *Process Innovation: Reengineering Work through Information Technology* (Boston: Harvard Business School Press).
Diamond, L. and Plattner, M. (eds) (1998) *Democracy in East Asia* (Baltimore, MA: Johns Hopkins University Press).
Friedman, T. (1999) *The Lexus and the Olive Tree: Understanding Globalization* (New York: Farrar, Straus & Giroux).
Gilbert, T. (1978) *Human competence* (New York: McGraw-Hill).
Herling, R. and Provo, J. (eds) (2000) *Strategic Perspectives on Knowledge, Competence, and Expertise* (San Francisco: Berrett-Koehler).
Jacobs, R. (1994) 'Comparing the Training Efficiency and Product Quality of Unstructured and Structured OJT', in J. Phillips (ed.), *The return on Investment in Human Resource Development: Cases on the Economic Benefits of HRD* (Alexandria, VA: American Society for Training and Development).
Jacobs, R. (1997) 'A Taxonomy of Employee Development: Toward an Organizational Culture of Expertise', in R. Torraco (ed.), *Proceedings of the 1997 Annual*

*Conference of the Academy of Human Resource Development* (Baton Rouge, LA: Academy of Human Resource Development) pp. 278–83.

Jacobs, R. and Jones, M. (1995) *Structured On-the-Job Training: Unleashing Employee Expertise in the Workplace* (San Francisco: Berrett-Koehler).

Jacobs, R., Jones, M. and Neil, S. (1992) 'A Case Study in Forecasting the Financial Benefits of Unstructured and Structured On-the-Job Training', *Human Resource Development Quarterly*, 3(2), pp. 133–9.

Jones, M. and Jacobs, R. (1997) 'Developing Front-Line Employees: A New Challenge for Achieving Organizational Effectiveness', in *The Guidebook for Performance Improvement: Working with Individuals and Organizations*, ed. R. Kaufman, S. Thiagarajan and P. MacGinnis (San Francisco: Pfeiffer) pp. 241–60.

Robinson, A. and Stern, S. (1997) *Corporate Creativity: How Innovation and Improvement Can Happen* (San Francisco: Berrett-Koehler).

Rummler, G. and Brache, A. (1995) *Improving Performance* (San Francisco: Jossey-Bass).

Swanson, R. A. (1996) *Analysis for Improving Performance: Tools for Diagnosing Organizations and Documenting Workplace Expertise* (San Francisco: Berrett-Koehler).

Watkins, K. and Marsick, V. (1994) *Sculpting the Learning Organization: Lessons in the Art and Science of System Change* (San Francisco: Jossey-Bass).

# Part II

# Human Intelligence Deployment in Asia

# 4 Employee Commitment: A Psycho-Social Perspective on Asian and American Culture and Businesses

William M. Czander and Dong Hwan Lee

## ORGANIZATIONAL COMMITMENT: PSYCHOANALYTIC PERSPECTIVE

The traditional view of the motivations for commitment to an organization has a strong base in economics. It is only recently that the research on employee commitment has included socio-cultural factors (Deal and Kennedy, 1982; Morgan, 1988; Kilman and Covin, 1998). The psychoanalytic view of employee commitment has a longer history. For example, Fromm (1970) suggests that the history of psychoanalysis, beginning with Freud, has dramatically rejected the economic view of 'system man' and has maintained a position that man's desire to seek and obtain commitment at work is not motivated by the need for material goods but by an array of social and psychological factors both conscious and unconscious.

The psychoanalytic view of commitment to an organization (Maslow, 1954; Hirschhorn, 1988; Czander, 1993) maintains that it is a function of interplay of the socio-cultural environment and the personality characteristics an employee brings to the corporation. Personality characteristics include both conscious and unconscious motivations which include wishes and fantasies. In addition the psychoanalytic position maintains that commitment results from a fit between these conscious and unconscious motivations, and the characteristics of the workplace environment. Consequently, psychanalysts view commitment to an organization as a psychic function, in that it serves both a defensive function (it reduces anxiety), and also functions as a vehicle for the gratification of conscious and unconscious wishes. Psychoanalysts see an employee's wish for commitment as universal, except in cases of psychological dysfunction or psychopathology. They see this motivation to be committed as an unconscious motivation, as a wish for human connectedness and a function of the drive to master, create, and accomplish. In the psychoanalytic literature this wish for human connect-edness is controversial. Some consider it a biological drive or instinct (Hendrick, 1943), others consider it an important part of the person's devel-opmental growth and relationships with the family of origin (Kohut, 1971; Czander, 1993).

There is widespread agreement among psychoanalysts that commitment is associated with occupational and career aspirations. It is well documented in the psychoanalytic literature that these aspirations evolve from the family dynamics and the person's socio-cultural environment (Hartmann, 1958, 1984; Kohut, 1971, 1985; Czander, 1993). If one follows this line of reasoning, then an employee's capacity to be committed to an organization evolves from a combination of family history, family dynamics, and the larger culture that influences the family. But what is the key linkage that connects this combination?

Romzek (1989) and others provide the linkage with their definition of organizational commitment. They define commitment as a 'sense of attachment to a work organization'.[1] Attachment is the conceptual link in understanding this relationship between the family, culture and commitment. Therefore, understanding attachment theory allows one to assess the relationship between the employee's past and the type of environment the organization provides, that will either deter or promote commitment.

## Attachment: The Link between the Employee and the Organization

From its origins in the psychoanalytic literature of developmental psychology, attachment theory has come to embrace many levels of meaning (Stern, 1985). Attachment theory is an attempt to make sense of the basic need for human connectedness; it is the end point of a developmental process involving the interplay of predefined and acquired behaviours. Attachment is typically used in the psychological sense as having an investment in, and an 'affective emotional component' to an object (Klein, 1964, 1975). In develop-mental theory, attachment occupies an important place in understanding the child's (and later on the adult's) capacity to form rewarding and stable relationships. (Mahler, 1972; Mahler, Pine and Bergman, 1975). In this sense, attachment is viewed as a capacity – a capacity to form and maintain rela-tionships. The theory of attachment maintains that this capacity evolves from the child's family and the family's capacity to form and sustain attachments. The researchers on attachment theory suggest that breaks or disruptions in attachments that occur early in one's life, especially in the family, lead to a reduced capacity to form attachments later in life (Bowlby, 1960, 1969, 1973; Stern, 1985). A diminished capacity to form attachments is typically demonstrated in early school years, and if unchanged will stay with the person through adult work life.

Attachment theory is primarily a theory of the family and child development, but to understand commitment fully the larger culture must be included. For example, it is this larger culture that will influence the child's belief system concerning what is appropriate or inappropriate with respect to the forming and sustaining of primary attachments within any given family. This is also true with respect to forming attachments outside the family. In all cultures, membership in an organization (school, tribe, community or work) that takes on significance early in one's life (Masterson, 1981). The importance or the degree of attachment one maintains to these institutions is established early on in a child's life, and is influenced not only by the child's capacity to form an attachment but by the expectations and values given to attachment by the larger culture.

**Western Culture and Attachment**

As children grow and develop, their involvement in institutions increases. In all cultures a significant degree of dependence on institutions begins in the adolescent period. However, the degree to which institutions influence an adolescent's movement away from the family differs greatly from culture to culture.

In Western culture, and in particular American culture, it is in the adolescent period that the child begins the process of rejecting family life, and of leaving home. This process of leaving home generally begins in early adolescence, as one moves, with greater degrees of intensity, into the peer community. Some experts suggest that liberation from family authority begins with the acquisition of an automobile, and in some places this can happen at age fifteen. More recently experts suggest that the beginning of movement away from the family begins at an even earlier age, (as early as 12). Experts point to sleep-overs and sleep-outs away from the family home, and suggest these activities are as commonplace as summer camps or trips abroad. It is common for 14 and 15-year-old teenagers, in affluent communities, to take spring, winter or summer unsupervised vacations with peers. In many regions children aged 12–14 have their own credit cards, and spend large amounts of time at malls where they eat out and shop in their own stores. These so-called 'mall rats' spend their weekends at the mall, a teenage haven, and continue their congregating once the mall closes in sleep-overs away from the family. Emancipation from the family continues when these teenagers reach 16 when they are able to drive automobiles and go to work at part-time jobs (usually in the malls). Liberation from the family increases as the teenagers now have their own money. It is commonplace for a teenager from a middle-class family to not only have his or her own car, but also their own televison, telephone and computer, which allows him or her to expand the numbers of hours embedded in his or her own subculture. Teenage television shows and teen-targeted advertising typically stress activities away from family, and in most instances the extended family is completely absent.

The process of leaving home ends when the teenager leaves for college, an almost universally accepted process in American culture. The process of leaving home is completed upon graduation from college and with entry into the organization (going to work). Stated simply, in American culture, the process of leaving home begins in the early teenage years and ends in the late teenage years or early adulthood. If the teenager has not left home in their late teenage years or by early adulthood, this is looked upon as problematic. From a cultural perspective, leaving home is uniquely consistent with the perception of economic success, or of attaining financial independence. The motivation to leave home is so strong in American culture, that for a young adult not to leave home may even be associated with shame and considered a function of some type of psychological problems having to do with dependence and autonomy issues. It is quite common for parents to react to the young adult who wishes to stay at home by forcing him or her to leave.

## Asian Culture and Attachment

We obtain a very different picture of the relationship between culture and commitment when we examine Asian culture. For example, the process of emancipation from parents, the rejection of parental authority, is not nearly as apparent in Asian cultures. In Asian cultures, adolescents are typically tied to their families economically and socially. Economically they are expected to contribute their paid and unpaid labour to the family system.[2]

In most Asian countries, children remain in the family throughout the adolescent period and into adulthood. They live with parents while they attend college, and even after college young adults live with their parents until they get married. There are two main reasons for this: cultural and socioeconomic. The cultural and economic garnishments available in Western cultures to expedite an early departure from the family, are clearly absent in most Asian societies. For example, mobility, a powerful characteristic found in American culture, is far less prevalent in Asian countries. This can be understandable because extended family members tend to live in close geographic proximity. The root of immobility in Asian countries is traced back to their agricultural society where family members collectively provided scarce labour to their farming business. In Asia, agriculture used to be the dominant industry throughout its history and is still a respected virtuous occupation. This is one of the primary reasons why the Japanese still fiercely protect their inefficient agricultural sector from global competition. The lack of mobility can be understood as one of the important contributors to commitment or having an attachment to a company. Asian workers are reluctant to leave a company, especially if it requires moving to another region. In contrast, American workers are not only willing to move when necessary, they typically value it. The metaphor 'Don't fence me in' has been valued since the pioneers days of early settlement in the American west. When Americans form attachments they are very different in quality from Asian attachments, which will be examined later.

**How Organizations Induce Commitment: Comparing Asian and American Corporations**

If we were to compare Asian corporations with their American counterparts we would observe several distinct differences concerning their structural nature. From a relative viewpoint the organizational structure of Asian firms tends to be more 'vertical', whereas that of American firms is horizontal. In Asian corporations, membership is based on a network of relationships that binds individuals into a cohesive group. The network of relationships can be 'familial' (extended family concept), 'regional' (from the same province), or 'educational' (graduates of the same school). When Asian companies hire employees, they focus more on overall qualifications, intensity of loyalty, and long-term commitment than on specific attributes or skills. Asian companies tend to believe that specific skills can be taught (and hence learned) as needed. Since employees are hired for life, both the company and the employee are willing to invest time and resources for training (and learning).

Compare the above approaches with those of Western companies. Employees are selected because their attributes and skills can serve the company's needs. Therefore, when the company no longer needs an employee's particular attributes or skills, there ceases to be a recognized basis for the worker's employment in the company. For example, if a worker's skill is in precision metal processing, when the company decides to move out of the metal processing business and go into the home electronics business, the worker is no longer needed.

However, in Asian companies, the membership is not necessarily based on skills and attributes, but rather on relationship and commitment. Since the company's goal and strategy may change over time, a long-time employee's skill may not meet the company's newly emerged needs. However, such a change is less important as a basis for membership in the organization. Using the previous example, when the company goes out of the metal processing business, the company is still committed to keep the employee whose main skill is processing metals, and will train the worker in the home electronics business. Although the adjustment process may take a long time, and sometimes turns out to be inefficient in an economic sense, the company believes that the worker is still valuable as long as he is committed to the company. As in the family, one of the goals of the company is the 'welfare' of its employees. This approach is consistent with Confucius' teachings, the philosophical underpinnings of attachment and commitment (this is described below).

From this perspective it is not difficult to see why an Asian employee's attachment to a company is 'total', and why commitment and loyalty is much stronger than any form found in American corporations. We can also see that the dedication and trust are 'mutual' in this vertical relational structure. This type of structure promotes the type of attachments necessary to strengthen commitment. They are so strong that Asian employees believe that they should stay with their company even when their company experiences financial difficulties. Under these conditions Asian employees are often more willing to forgo compensation if necessary for the company's survival. This type of commitment is almost unheard of in American companies.

The hierarchical structure of the Asian corporation resembles the family. In the Asian culture, the family, above all else, must be given respect and honour. Senior family members obtain the highest level of respect and honour. Adolescents are required to consider all their actions, with family reverence and respect in mind; this includes educational success, vocational choice, and choice of friends and romantic involvements.

Independence is not a cherished value in Oriental culture, both in the family and the organization. Being 'in' is more valued than being 'out', a value more prevalent in American culture. Being 'in' means being in the group. The priority is always on the group over the individual, and consensus and harmony are paramount concerns of social relationships.

## Vertical and Horizontal Structures: The Cultural Aspects of Authority and Order

The Asian organization is marked by two rigid characteristics: (1) its vertical structure and (2) the demand for order. The hierarchical arrangement where the paternal figure occupies the top position in the family is reflected in the Asian organization. The pervasiveness of hierarchy also exists among siblings in the family where a strict order is maintained. Older siblings are respected and have authority over younger siblings (similar to parental authority). The oldest son is trained from an early age to assume the leadership when the father dies. For example, it is taboo that a younger brother marries before the older one gets married. This kind of order can be reversed only with the family approval. We find no evidence of this demand for order with American families. Even the paternalistic and hierarchical structure in the family is disappearing. Consequently, demands for authority-based hierarchies and bureaucratic order are rapidly becoming dinosaurs in corporate America. Such a trend is evident from the fact that the MBAs graduating from elitist schools avoid these hierarchical organizations, and instead seek out what is rapidly becoming the American ideal, the horizontal, loosely structured organizations, where rapid growth and risk-taking are apparent.

There are three aspects of American culture that work against the establishment of employee commitment. First, the weak psychological sense of community in American culture and the lack of stability and often avoidance of 'attachments' within the family and community. Second, the employee's sense of power and value to move against authority. If one traces American corporate history, it is this move against paternal authority that is valued and even celebrated. It is paternal authority that is perceived with mistrust. Whereas respect towards authority is well founded in Asian culture, in American culture the negative view of authority is well established. This negative attitude toward authority is embedded in American history beginning with the revolt against monarchies, perceived as objects that deny individual freedoms and liberties. In corporate America, this revolt is firmly represented by the history of the union movement and management–labour strife that still exists today. In American culture submissiveness and deference to authority is experienced as demeaning, and self-respect is attained through competition with others, or against the larger 'system'. Behaviours are expected and highly valued if they emphasize the employee's uniqueness or separateness (Morgan, 1985). On the other hand, in Asian companies, deference to authority is expected and harmony and unity are valued. Speaking up against authority, a value embedded in American culture, holds a powerful stigma in Asian culture, especially if someone should lose his or her job because they spoke up. If a problem exists at work, Asian employees will avoid direct confrontation (an approach prevalent in American corporations). Rather, they tend to take a circumspect, indirect way of dealing with management. Because Asian employees put the welfare of the group (company) before their own, they try to avoid disrupting the continuity and unity of the company whenever possible. They believe any confrontational issues should be solved with management in consensus, and consequently they are far more patient than Westerners in working out mutually acceptable solutions. Third, a value system that works against commitment, the value of the 'pioneer' spirit that has permeated all aspects of American life: the rugged individualism of the early settlers, the glorification of the entrepreneur, and the esteem given to American 'stars' or heros push against commitment to an organization. This is a culture that promotes individualism in contrast to the Asian spirit which values the collective cooperative spirit.

## Organizational Commitment in Asian and American Corporations

As discussed in the previous section the Asian culture clearly supports commitment to the organization. Commitment can be established with relative ease in a culture where a strong sense of community exists, stability is valued, and duty, respect, and honour predominate as an integral part of the employee's experience. But most important in establishing commitment is the powerful sense of obligation. In Asia this sense of obligation is 'mutual', between employer and employee (group). This mutuality is embedded in Asian philosophy, in particular, the teachings of the Chinese philosopher and sage, Confucius (551–479 BC). Confucius' teaching has had a tremendous impact on the values and moral fabric of the people, the culture, and the structure of social relationships in Asia throughout history. Confucianism is actually more of a system of ethics and morals than religion *per se*, and it stresses the obligations of people to one another and to social institutions as well as the relationship among them. Confucius advocated such virtues as sense of duty, loyalty, honour, filial piety, trust, respect for seniority, respect for authority, and sincerity, harmony and peace. It should be noted that Confucius' precepts emphasize social order and harmony in the 'vertically' (i.e., hierarchically) structured system. It is essential to understand not only the nature of the philosophy but also how a philosophy is communicated and accepted, an important aspect of creating commitment. Ouchi (1981) bears this out when he explores the difference between American and Japanese managers. He points out American managers are enamoured with hard facts and data. For example, when they state their company's objectives, they do so with impressive displays of data and facts. In contrast the Japanese allow things like objectives to evolve out of the processes associated with exploration and understanding of the values through which the company is or should be operating. For Asian managers, understanding and accepting the basic philosophy of the company is more important than hard data; it is the philosophy which functions as the guiding light from which objectives flow.

## The Psychology of Attachment: The Two Cultures Compared

Another way to assess commitment is to observe how these two cultures value and sustain the idea of being in a community in the psychological sense.

In American culture one's membership in a community is based on common-alities such as age, race, ethnicity, education, work, physical and now electronic proximity. These communities tend to be temporary in nature. Because of the high degree of mobility and horizontal nature of human relationships, strong ties and attachments are often not stabilized. A common example is the teenager leaving home for college. Not only young people but also adults move with ease from community to community, especially with educational needs, employment changes and shifts in economic status. Consequently, given the assumption that one will not remain long, relationships may be formed only if advantageous. We refer to the companies and communities where these types of relationships are formed, as 'narcissistic communities'.

In these narcissistic communities employees expect to extract gratification quickly, and to have their needs met; otherwise they move on. This culture of narcissism (e.g. Lasch, 1979; Rothstein, 1980) is typified by a 'me first' attitude, where one will work hard in order to be admired, and for the purpose of 'getting more' for oneself, even if it is at the expense of others. Consequently, these narcissistic communities become the arenas of instant gratification. This work-hard, play-hard, me-first attitude has been prevalent for a such a long time that American managers are resigned to it. They readily accept employees who cannot demonstrate more than minimal commitment, are quite mobile, and have an exceedingly low tolerance for delaying grat-ification. American managers in conjunction with consultants have learned to exploit this attitude. For example, in American corporations a young, just out of college employee will change jobs perhaps three times in the first five years of employment. American corporations encourage job changing, and it is rewarded. This is indicated by figures suggesting that job changing is a critical factor in increasing an employee's income. In America, with every job change the average increase in income is 12 per cent, while the average per annum increase is only 4 per cent. Employment agencies, so-called head-hunters, and recruiters are almost a major industry in America, suggesting that more and more employees are on the move. This trend is reinforced by the American corporations that value temporary employees, outsourcing, consulting, downsizing, and the recent proliferation of 'start-ups', extraordin-arily high-risk endeavours.

The fact that this type of instant narcissistic gratification culture is not apparent in Asian corporations is clearly a reflection of the larger culture. For example, in the Asian culture mobility is not integrated into the cultural fabric because the family takes on a significantly greater degree of involve-ment in one's psychological sense of community. In the psychological sense this community is protective and defensive; that is, the family community prevents the loss of integration that one experiences when isolated. The function of this type of stability is not to provide narcissistic gratification but to ward off and protect the person against real or imagined 'narcissistic injuries', especially the injury associated with being rejected and alone.

Consequently, we see a clear contrast of two styles of narcissistic pursuit: one (American) associated with the conscious and unconscious wish of getting more for oneself, of being seen by others as perfect and the other (Asian) associated with avoidance of the injury associated with shame, rejection and isolation.

In the latter, attachment is a deep and powerful concern and plays an important role in the employee's narcissistic pursuits. In the former, attachment serves a different utility, and is easily broken if the corporation fails to offer narcissistic gratification. We will discuss next these different styles of attachment played out in the form of transference reactions to the organization.


## THE DYNAMICS OF ORGANIZATIONAL TRANSFERENCE AND COMMITMENT: A PSYCHOANALYTIC PERSPECTIVE

In everyday life rarely, if ever, is any relationship free from transference reactions (Czander, 1993). Beginning with Freud (1958/1912), psychoanalysts have claimed that transference permeates all perceptions and communicative processes. Durkin (1964) confirms the position that transferences evolve from social interactions that in groups and organisations transferences collide, meet and fuse.

All forms of commitment require what are referred to as *transference reactions*. Although the psychoanalytic literature abounds with research on transference in the clinical sphere, little attention has been paid to transference as it occurs in the organization. Greenson (1967), following Freud, explored the pervasiveness of transference and implied that transference reactions could be applied to institutions. He suggested that while transference reactions were essentially unconscious, that is beyond awareness, some aspects may be conscious. For example, a person may be aware that he or she is reacting strangely or with excessive anxiety to a particular situation, although that person may be unaware of the meanings or motivation for this strange reaction. Also, a person may be intellectually aware of the source of his or her reaction to a particular environment, but the emotional basis that motivates the reaction may be beyond awareness. Consequently, a transference reaction, either positive or negative, occurs when a person consciously or unconsciously projects earlier wishes, feelings or fantasies on to an object. In this sense, transference reactions are repetitions of these early experiences.

These reactions are interactive, which means that the organization is capable of inducing transference reactions.

From a psychoanalytic perspective, organizational commitment requires elements of a positive transference reaction. In the organization, commitment results when the employee is capable of displacing or transferring positive feelings and attitudes on to the organization. In a historical sense, these feelings that are transferred to the organization evolve from some significant person(s), objects or institutions in the employee's past. These objects from the past that influence the present can be parents, siblings and teachers, but also inanimate objects like buildings or other objects in their physical environment. This explains the importance of culture and early family dynamics. It is these dynamics that will influence the employee's later capacities for forming positive transference reactions. Consider the following case of an employee who has chosen truck driving as a vocation. This person has experienced conflicts with authority his entire life: problems at home, problems in school. He was an early school dropout, and this culminated in an occupational choice, driving a truck. In this occupation he shows up for work, engages in conflict with his boss (authority), and then he spends his day free from the psychic oppression of authority as he drives from location to location delivering his 'goods'. At the end of the day he goes back to the office, for another brief encounter with the boss. For this person to work all day in an authority relationship would be exceedingly problematic; his early conflicts with authority in the family and school would be repeated. His history of conflict relationships with authority proves fertile ground for later negative transference reactions. He cannot form a positive transference, and consequently he shows little commitment to the company he works for. He is frequently late for work, he quickly consumes his sick days and personal days, and if it were not for the union that represents him, his employment most likely would have been terminated.

From a psychoanalytic perspective all transference reactions, including positive transferences (commitment), if they are to be sustained, must be shared perceptions. These shared perceptions can be both good and bad. Also, they are a powerful dynamic that can promote bonding among group members; that is, if the perceptions are collectively shared, cohesion will result. Returning to the truck driver, if he shares his perceptions of the 'bad' authority with his fellow truck drivers, and they collectively concur about authority, then they will bond together in their mutual feelings. As Durkin (1964) suggests, shared perception is the dynamic underlying commitment to an organization; that is, commitment stands or falls within the context of a group. Another way of observing commitment in the context of the group is the employee's experience of the organization as omnipotent. When employees collectively identify with this power – that is, they take it in, introject it, and make it their own – then together they feel powerful. Commitment is sustained as long as this identification is sustained; the employees feel power. It is a shared perception when other employees collectively identify with the many attributes of the organization.[3] This view of an identification with certain characteristics of the organization and not the entire organization may be unique to American corporations. In many American corporations employees may be commited to a CEO (leader), and not to the organization. This occurs because executives in American corporations hold the position that it is the job of management to induce commitment.

**American View of Inducing Commitment in Employees**

In American corporations commitment is leader-oriented. The belief that the leader is expected to instil a sense of commitment in employees has its roots in the structure of the American military. In the military, all recruits must undergo a socialization process, 'boot camp'. The purpose of the socialization process is to induce in the new recruit a powerful sense of commitment to the degree that the recruit will unquestionably give his or her life for the 'cause'. This socialization process instils in the recruit a lack of self-identity, and over a period of many days with intense drilling and physically and psychically exhausting exercises, the recruit gives up certain mental and physical capacities. In the case of the military, the recruit gives up the capacity to think. The group, the collective unit of recruits, bonds together as a committed group. This occurs psychically when they all project on to their platoon leader the capacity to think, and then the recruit identifies with that which they projected. In this case, the leader thinks and the soldiers follow *en masse*, with collective obedience. This is referred to as a powerful positive transference, which forms the psychic base for commitment that evolves when the recruits, who have collectively projected a common capacity on to the leader, identify with that capacity (the leader is brilliant). Over time, the recruits bond together when they collectively value the defence provided by a 'narcissistic identification' with the leader, a defence against feelings of inferiority and vulnerability associated with the task of the military, which is to fight. As stated above, American theorists typically assume that behaviours associated with commitment occur as a function of the presence of a type of leadership (Burns, 1978; Meindl, Ehrlich and Dukeric, 1985). However, these commitment behaviours associated exclusively with leadership are artificial, in that they lack genuine enthusiasm and are rarely displayed in the absence of the leader. Freud (1955/1921) articulated this in his study of the war between the Assyrians and the Hellecians. When Judith lost his head in battle (was decapitated), his troops behaved as if they had lost their heads. In spite of ample evidence suggesting that leader-based commitment is wrought with problems, it remains popular in American management culture and is amply reinforced by management gurus and consultants who maintain that leaders are not only capable of inducing commitment, but should (Peters, 1992). Seminars, lectures, and consultations on how to induce a 'good enough' environment by which a family can nurture and develop a child. commitment are popular because they allow the continuation of traditional ideas about commitment that are unique to American corporate culture, especially ideas associated with the glorification of power and leadership. This is especially apparent when one considers the view of commitment that has existed in American literature since the beginning of the grand migrations of immigrants and the industrial revolution (see Taylor, 1911; Braverman, 1974). Recognizing the harsh history of problematic management–labour relations and the vast mobility of the population in America, the belief prevails that the larger culture does not support commitment, and may actually move against it. Consequently, American theorists and corporate leaders have taken it upon themselves to embrace the belief that it is management's duty to induce and, if necessary, use economic and psychological manipulations to force commitment in employees. To a degree this makes sense, especially if one observes the corporation as a narcissistic community containing employees who are preoccupied with their own advancement. In this case, commitment among these employees is so weakened that it can only be attained through a (narcissistic) identification with the power of leadership (Peters, 1992; Kanter, 1968). This may explain the glorification of CEOs in American culture, where they become wealthy media celebrities.

In summary, American theorists rely exclusively on leadership when putting forth ways to induce commitment in employees. They typically propose five ways to induce commitment:

1. Through hierarchical arrangements, authority structure and role requirements, and task arrangements. Durkin (1964) and Halpern and Halpern (1983) claim that in organizations, transference reactions are precipitated by the structure, and it is structure which provides communicative channels (status differential, proximity, and roles).
2. Through the organization's culture, history, rituals, customs and norms, and its inanimate objects, like computers, office space, furniture, the architecture, and technology, etc. For example, re-engineering writers (Hammer and Champy, 1995) and cultural writers Peters and Waterman (1982) believe that commitment can be created by manipulating these artifacts.

3.  Through the organization's demands for performance, duty, and task requirements, and through dispensing of rewards and punishments (Komaki *et al.*, 1996; Mowday, 1996).
4.  Communicating a company's shared vision and establishing a shared mission with employees are important means of enhancing the employee' motivation and commitment (e.g., the vision statement and the mission statement) (Drucker, 1973).
5.  Finally, the most consistently popular method of inducing commitment (as indicated in the military example) suggests that it will evolve when the employees identify with their leaders. Therefore, employees must see their leaders as all-powerful, and consequently, the corporation's leadership surround themselves with an array of symbols associated with power (large salaries, staff, opulent offices, home, furnishings, limousines, etc.).

**Commitment and the Asian Corporation**

Psychoanalytic writers believe that if the organization is to induce commitment a form of regression[4] is necessary (Czander, 1993, 1995; Reider, 1953, 1957). They see regression as capable of promoting identification with leaders under certain conditions, but they widen the field to include regressions that can promote identifications which more uniquely fit the Asian culture. For example, commitment evolves in the Asian culture when the corporation comes to serve as a haven or refuge, where employees feel a sense of security and a maternal sort of protection against the anxiety associated with rejection, shame and isolation. In response to this anxiety, the organization and the employees, peers and colleagues, take on a maternal function. The corporation induces commitment when it is capable of filling the employee's need for security, and not power, as it would in the American corporation. When we think about the psychic function that commitment serves in the Asian culture, it provides a protective front in the sense that it helps the employee cope with vulnerabilities, insecurities, and narcissistic injuries. This form of regression fits with Bowlby's (1969) view of the 'good enough' environment by which a family can nurture and develop a child.

**Psychoanalysis of the Corporation as Negative Transference Object or Negative Commitment**

If the corporation is unable to create a 'good enough' environment and if the employee must daily work within a competitive, fearful environment where narcissistic injuries prevail, the employee will be unable to form narcissistic identifications with the organization. Consequently, the employee's ego will not be able to sustain a constant object relationship with the organization. This will eventually culminate in a negative transference reaction (the opposite of commitment), a phenomenon that is frequently found in American corporations. These organizations are experienced as a lost object, which results in a loss of confidence, a loss of interest, and the re-experiencing of affects associated with anger and aggression. On the other hand, employees can experience a negative transference and form a powerful commitment to the organization in their unity against the organization, namely its leadership. For example, shared aggression, in particular, is important because it is a way in which one can bring the object (organization) close to experience and capture a union, an attachment. The attachment results from the shared hatred of the corporation. In this instance the organization is no longer the admired object and aggression and hostility replace the previously wished-for infantile magical attempt at a union with the object. The organization becomes the opposite of a maternal haven. When this occurs the employee moves to one of three positions:

1.  The first position is where he or she withdraws from the emotional life of the organization and functions more or less in a 'robot' manner on the job, constantly thinking about and fantasizing about a better place to work.
2.  In the second position the employee views the organization as a game and thinks of survival in this game as defeating real or imagined enemies within the organization to get ahead, obtain more status, and become powerful like the leadership. The goal of winning becomes a displacement for the hatred the employee feels for the organization and its leadership, that is, the employee displaces his or her aggression. In these types of ruthless, competitive corporations, the 'empathic net' is missing, a 'holding environment' is lacking and the employees frequently report they are similar to a high-wire act, performing without a net. In this harsh work environment, the employee can only receive imagined gratifications (narcissistic supplies) by engaging in self-serving, aggressive, and competitive behaviours. In these typically American corporations we see the virtues of Western culture at play: the notions of individualism, being a star, being better than others; and the use of metaphors like 'the cream rises to the top'. Consequently, in these unemphatic corporations, commitment is not apparent, and management either consciously or unconsciously rejects this idea as inconsistent with their so-called macho culture. In these corporations the emphasis is on working hard, learning, resumé building,

making money and the constant search for something better. The employee is not only permitted but is encouraged to compete with fellow employees. As a matter of fact, we find an inverse relationship between corporations that devalue commitment and corporations that value intense competition.

3. Finally, the employee maintains commitment through an identification with the aggressor (corporation and its leader). This employee wants to be like them, and he or she takes on their characteristics, feels fear in their presence and is enamoured with the wish and resulting fantasy of obtaining what they have. The striving is unconsciously motivated by the wish to be the favourite son or daughter. Obedience is a cover for the underlying aggression to consume the idealized leader.

When one looks at organizations that have created holding environments, and have successfully established sanctions against competition, aggressive behaviours, and individualism (typical Asian organizations), they are able to promote and maintain a positive transference (strong sense of commitment) among their employees. In these organizations, the employee can identify with the organization as a safe, maternal harbour. The illusion that the organization is powerful does not create fearful responses. In this arrangement it is through identifications with the organization that the employee experiences a sense of competence, and not the envy described above.

## The Ideal: The Psychoanalysis of Genuine Commitment

In the larger Asian culture where there exists a strong psychological sense of community and harmony and powerful family bonds, we find an increased capacity for shared narcissistic identifications. When shared narcissistic identifications are operating, employees show diminished investment in the need to idealize the leader. The employees as a group take on the characteristics associated with an 'all business' attitude. In addition, movements against the organization, comments critical of the organization and its leadership, are de-emphasized. Only within a group of genuinely committed employees can conflict and struggle to meet objectives and goals be tolerated and transformed into productive work. Shared narcissistic identifications provide for creation of an emotional tie, and it is the strength of this tie that permits struggle and conflict without the fear of authority, fear typically associated with quasi-committed organizations. Transference reactions that evolve from these types of narcissistic identifications organizing activities are restorative in a psychological sense. For these powerful reactions to occur the corporation must be able to do the following:

1. gratify aspects or derivatives of the employee's core or golden fantasy. This is accomplished when fantasies associated with authority are gratified and these fantasies collectively evolve when shared by a group. When this occurs, the employees in this group have an increased capacity to control their own behaviour and wishes, utilize moral restraint and engage in self-punishment.
2. promote a comfort zone within the corporation where employees can create their own reality or fantasy to adapt to difficult realities just as their family does. This allows the corporation and their formal and informal work groups to maintain and/or restore precarious, disintegration-prone self and object images.
3. provide and strengthen necessary defences to ward off configurations of experience which are felt to be conflictual or dangerous.
4. provide to the employee a hope that he or she will be valued and that the corporation will provide the same haven and sense of community in an ideal sense that is experienced within the family. The organization is seen as capable of buffering the employee from the dangers associated with the external world, and his or her own internal reality, just as family and community have done in the past. To do this, the organization must function as a maternal haven by being constant, reliable, non-critical, and empathic, and at the same time show patience and set limits.
5. establish leader–followers relations that are not feared and criticized, not harsh and punitive, but held in high esteem, respected and valued.

When the corporation fails to provide a holding environment, as many American corporations do, and at the same time demand that the employee abandon real and wished-for attachments by creating a competitive and judgmental environment, the corporation will diminish the motivation for commitment and instead promote depressive and persecutory responses. On the other hand, when the organization provides an 'empathic net', the employee responds to the organization as a positive object with a heightened sense of commitment. When this occurs, the organization takes on a force in the employee's life similar to that of the family, and the wish for this commitment continues over his or her life course (Ingber, 1981). These employees do not wish to be away from the organization and, even when they are separated, they find ways to engage in an object relationship with it. Their identification with the organization is symbolically communicated through their dress (wearing company logos), and their proud display of membership. When they meet strangers they introduce themselves with their company card, which is not only a way of communicating and confirming their identity, but expresses the wish of potential continued cooperation with the stranger in their protected role as a corporate representative.

Committed employees use instruments or objects (like laptop computers, beepers, or cell phones) as symbols, that is, as symbolic attachments or symbolic cords that allow for a continuous attachment to the organization. In these cases, these instruments assume the role of an object capable of performing Bion's (1970) 'containing' function and Winnicott's (1958) 'holding' function. This results in an attachment that performs the same psychic function as the family, which gives the employee a sense of comfort and security. Commitment viewed from this perspective is the continuous experience of a union between the employee and the corporation, where the employee in fantasy can merge with the gratifying object, the corporation.

## CONCLUSION

A comparative analysis of American and Asian cultures and how families from these respective cultures support or negate attachments, will contribute to our understanding of the concept of employee commitment. It is expected that this comparative analysis will offer human resource managers and others concerned with organizational commitment a clearer understanding of how the larger culture influences the family and how these capacities to form attachments in the family will contribute to, and/or detract from, an employee's commitment to the organization. It is hoped that this type of assessment will give human resource managers an opportunity to assess cultural changes, in particular how cultural change within Asian families and the larger community will impact employee commitment over time. In particular, it is suggested that as Asian culture and family are increasingly influenced by Western thought and behaviour, organizational commitment will become less apparent in Asian corporations.

### Notes

1.  Although there are different views and definitions of commitment in the literature, reviewing them is beyond the scope and the intent of this chapter. Interested readers, however, are referred to Ghemawat, 1991 (commitment as a form of persistence), Kaplan and Razin, 1981 (commitment as a strong attachment to an organization), and Ingram, 1986 (commitment as a sense of intimacy).

    While all of these views enhance our understanding of the concept of commitment, they do not explain the depth of the relationship that is established when an employee becomes committed. For example, what it is about the object that receives commitment, and what is it about the committed person and their relationship that provides a powerful sense of commitment? This is a major question asked by organizational theorists and managers alike as they attempt to promote and reap the benefits associated with commitment. On the other side, the critical questions associated with the concept of commitment are What effect does it have on employees? Does it provide for increased mental health, and job satisfaction? Is it an ideal condition, or is it nothing more than a form of psychic bondage where employees are brainwashed into demonstrating the same selfless type of commitment that young adults demonstrate when they join a cult and sell flowers at an airport all day for a bowl of rice and praise from an over-idealized guru?

2.  Minor exceptions can be found within Malaysian or Javanese cultures where youth are reluctant to submit to parental authority at least in an economic sense (Hefner, 1998).

3.  This is not to dismiss the significance of the employee's inner and earlier experiences and the notion that transference reactions are repetitions of earlier experiences. For example, some employees will tend to react negatively and show little response to the organization's desire to induce a positive transference reaction. In addition, the quality of the organization's psycho-social environment may be of little consequence in inducing a transference reaction because the organizational environment will attract personality types that will find the climate appealing. The key here is to match the personality of the recruits with the psycho-social environment. For example, research has suggested that if the personality of an employee is unable to fit into the modal personality or culture of the organization, a consequence will be increased stress (Kets de Vries and Miller, 1984; Czander, 1993). An example of this type of stress was found with an automobile manufacturing employee who had to quit his job because of a mental breakdown, and the courts ruled that the breakdown was the result of an incompatibility between his personality, his type of job, and the culture of the factory. He was a parts inspector in the assembly line and he suffered because the workers kept installing defective parts on the assembled autos. His personality demanded order, correctness, and cleanliness; he would be diagnosed as an obsessive compulsive personality. On the job he experienced dread. Working within a culture that permitted messiness and defects had a direct bearing on his mental health. In all cases where an employee demonstrates personality characteristics that do not fit in to the larger culture, this employee will be perceived as being out of step, as different, and will most likely experience stress. If the parts inspector were able to manifest and gratify his obsessional zeal, he might have been a contented and committed employee, and the organization as a whole may have benefited.

4.  Regression as it is used here is a move away from higher levels of mental functioning to more archaic modes. It is not used in the sense most often used in the psychoanalytic literature, as a return to some earlier period of childhood or a re-enactment of some childhood trauma. It is not used as a return in the temporal sense.

## References


Bion, W. (1970) *Attention and Interpretation* (New York: Basic Books).
Bowlby, J. (1960) 'Separation Anxiety', *International Journal of Psychoanalysis*, 43, pp. 89–113.
Bowlby, J. (1969) *Attachment and Loss. vol. 1 Attachment* (New York: Basic Books).
Bowlby, J. (1973) *Separation: Anxiety and Anger: vol. 2. Attachment and Loss* (New York: Basic Books).
Braverman, H. (1974) *Labor and Monopoly Capital* (New York: Monthly Review Press).
Burns, J. M. (1978) *Leadership* (New York: Harper and Row).
Czander, W. M. (1993) *The Psychodynamics of Work and Organizations* (New York: Guilford Press).
Czander, W. M. (1996) 'The sick building syndrome: A Psychoanalytic Perspective', *Forum in Psychoanalysis*, 12, pp. 28–39.
Deal, T. E. and Kennedy, A. A. (1982) *Corporate Cultures* (Reading, MA: Addison-Wesley).
Drucker, P. (1973) *Management: Tasks, Responsibilities and Practices* (New York: Harper and Row).
Durkin, H. (1964) *The Group in Depth* (New York: International University Press).
Eagle, M. (1981) 'Interests as Object Relations', *Psychoanalysis and Contemporary Thought*, 4(4), pp. 527–65.
Farber, L. (1976) *Lying, Despair, Jealously, Envy, Sex, Suicide, Drugs, and the Good Life* (New York: Basic Books).
Freud, S. (1955/1921) *Group Psychology and the Analysis of the Ego*, in J. Strachey (ed. and trans.), *The Standard Edition of the Complete Psychological Works of Sigmund Freud* (vol.18, pp. 67–143) (London: Hogarth Press) (original work published 1921).
Freud, S. (1958/1912) *The Dynamics of Transference*, in J. Strachey (ed. and trans.), *The Standard Edition of the Complete Psychological Works of Sigmund Freud* (vol. 12, pp. 97–108) (London: Hogarth Press) (original work published 1912).
Fromm, E. (1970) 'Thoughts on bureaucracy', *Management Science*, 16(12), B699–B705.
Ghemawat, P. (1991) *Commitment: The Dynamics of Strategy* (New York: The Free Press).
Greenberg, J. R. and Mitchell, S. A. (1983) *Object Relations in Psychoanalytic Theory* (Cambridge, MA: Harvard University Press).
Greenson, R. (1967) *The Technique and Practice of Psychoanalysis* (New York: International Universities Press).
Guntrip, H. (1969) *Schizoid Phenomena, Object Relations and the Self* (New York: International Universities Press).
Halpern, J. and Halpern, I. (1983) *Projections* (New York: Seaview/Putnam Press).
Hammer, M. and Champy, J. (1995) *Reengineering the Corporation* (New York: Harper).
Hartmann, H. (1958) *Ego Psychology and the Problem of Adaptation* (New York: International Universities Press).
Hartmann, H. (1984) *Essays on Ego Psychology* (New York: International Universities Press).

Hefner, R. W. (ed.) (1998) *Market Cultures* (New York: Westview Press, Columbia Institute of Southeast Asian Studies).

Hendrik, I. (1943) 'Work and the pleasure principle', *Psychoanalytic Quarterly*, 12, p. 311–329.

Hirschhorn, L. (1988) *The Workplace Within* (Cambridge, MA: MIT Press).

Ingber, D. (1981) 'Computer addicts', *Science Digest*, July, 114–121.

Ingram, D. H. (1986) 'Remarks on intimacy and the fear of commitment', *American Journal of Psychoanalysis*, 46(1), pp. 76–79.

Kanter, R. M. (1968) 'Commitment and social organization: A study of commitment mechanisms in utopian communities', *American Sociological Review*, 33, pp. 499–517.

Kanter, R. M. (1983) *The Change Masters* (New York: Simon & Schuster).

Kaplan, S. R. and Razin, A. M. (1981) 'Psychosocial dynamics of group cohesion', in H. Kellerman (ed.), *Group Cohesion* (New York: Grune & Stratten).

Kets de Vries, M. F. and Miller, D. (1984). *The Neurotic Organization* (San Franscisco: Jossey-Bass).

Kilman, R. H. and Covin, T. J. (1988) *Corporate Transformation* (San Francisco: Jossey-Bass).

Klein, M. (1964) *Contributions to Psychoanalysis, 1921–1945* (New York: McGraw-Hill).

Klein, M. (1975) *Envy and Gratitude and Other Works, 1946–1963* (New York: Delta Press).

Kohut, H. (1971) *The Analysis of Self* (New York: International Universities Press).

Komaki, J. L., Coombs, T. and Schepman, S. (1996) 'Motivational implications of reinforcement theory', in R. M. Steers, L. W. Porter and G. A. Bigley (eds), *Motivation and Leadership at Work*, 6th edn (New York: McGraw-Hill) pp. 35–52.

Lasch, C. (1979) *The Culture of Narcissism* (New York: Werner Books).

Mahler, M. (1972) 'Rapprochement sub-phase of the separation-individuation process', *Psychoanalytic Quarterly*, 41, pp. 487–506.

Mahler, M., Pine, F. and Bergman, A. (1975) *The Psychological Birth of the Human Infant* (New York: Basic Books).

Maslow, A. (1954) *Motivation and Personality* (New York: Harper & Row).

Masterson, J. (1981) *The Narcissistic and Borderline Disorder* (New York: Brunner/Mazel).

Meindl, J. R., Ehrlich, S. B. and Dukerich, J. M. (1985) 'The Romance of Leadership', *Administration Science Quarterly*, 30, pp. 78–102.

Mowday, R. T. (1996) 'Equity theory predictions of behaviour', in R. M. Steers, L. W. Porter and G. A. Bigley (eds) *Motivation and Leadership at Work*, 6th edn (New York: McGraw-Hill) pp. 53–72.

Morgan, G. (1985) *Images of Organizations* (Beverly Hills, CA: Sage).

Ouchi, W. G. (1981) *How American Business can meet the Japanese Challenge* (Reading, MA: Addison-Wesley).

Peters, T. (1992) *Liberation Management* (New York: Fawcet Columbine).

Peters, T. J. and Waterman, R. H. (1982) *In Search of Excellence* (New York: Harper & Row).

Reider, N. (1953) 'A type of transference to an institution', *Bulletin of the Menninger Clinic*, 7, pp. 58–63.

Reider, N. (1957) 'Transference Psychosis', *Journal of Hillside Hospital*, 6, pp. 131–49.

Romzek, B. (1989) 'Personal consequences of employee commitment', *Academy of Management Review*, 32, pp. 649–61.

Rothstein, A. (1980) *The Narcissistic Pursuit of Perfection* (New York: International University Press).

Stern, D. (1985) *The Interpersonal World of the Infant* (New York: Basic Books).

Taylor, F. W. (1911) *Principles of Scientific Management* (New York: Harper Row).

Winnicott, D. W. (1958) *Through Paediatrics to Psychoanalysis* (London: Hogarth Press).

Winnicott, D. W. (1965) 'Ego Distortions in Terms of True and False Self', in D. W. Winnicott (ed.) *Maturational processes and the facilitating environment* (New York: International Universities Press) pp. 140–54.

Wolberg, A. (1975) 'The Leader and Society', in Z. Liff (ed.), *The Leader in the Group* (New York: Aronson) pp. 247–50.

# 5 Psychosemantic Model of Social-Economic Behaviours: Multidisciplinary Considerations and Applications in the Asian Context

Oliver C. S. Tzeng

## MULTIDISCIPLINARY APPROACH TO SOCIAL AND ECONOMIC BEHAVIOURS

### International Business Scenario

In the contemporary global community, international communications and business exchanges are common and frequent phenomena, ranging from imports or exports of commercial goods to a joint venture in industrial productions. Assume that an American company plans to initiate a joint venture in manufacturing certain industrial products (e.g., auto parts) in Mainland China. The company plans to send a group of technical delegates to China for supervising all areas of production activities. Assume also that these American delegates have never been in China, and have no prior experience in working with Chinese workers. Thus, although quite excited about the new overseas assignments, they are extremely apprehensive of culture shock – not knowing how to 'behave' appropriately in cross-cultural transitions (Oberg, 1960; Searle and Ward, 1990).

Assume further that the company has retained an American expert in China who has successfully worked with Chinese employees. One of the major tasks of this expert is to develop an orientation programme for the new American delegates. The training contents can, in theory, be divided into two broad categories: *technical* – about particular occupations – and *psychosocial* – about general life adjustments in a foreign country (Hannigan, 1990). Since these employees are experienced professionals in their respective technical assignments, the training will focus primarily on the psychosocial aspect of adjustments – both inside and outside their immediate working environment in China. All issues surrounding this hypothetical training programme are the topics of this chapter.

## Theoretical Frameworks

Tzeng (1993) developed a process-oriented theoretical framework called the 'Psychosemantic Process Paradigm' for the study of general social behaviours. The concept of 'psychosemantic process' postulates that all overt social behaviours are the manifestations of an individual's internal mental states. This functional relationship represents a continuing process in acquiring, adjusting and applying an individual's implicit meanings of external worlds.

Under this framework, the entire process of typical international economic behaviours is operationally categorized into ten structural units (called *model components*). The functional relationships of these components can then be used to develop a training programme for acculturation of foreign delegates in a new country. Before the model components are described, other fundamental concepts embedded in the model development are introduced below:

*Two types of cultures.* Each individual has acquired a repertoire of unique characteristics through heredity, development, and social learning. These characteristics serve as foundations for interactions with all – familiar and new – external environments. These characteristics and all external environments can generally be organized under two types of cultures: (1) *Objective culture* represents the characteristics that are readily observable and generally measurable (such as income level). (2) *Subjective culture* represents the internal conditions that are not directly observable or measurable (such as attitudes and beliefs). These two cultures constantly interact and reciprocally reinforce in our daily lives.

*Psychosemantic process.* Humans constantly encounter a variety of new 'objects' (people and situations) that require us to interpret the objects, evaluate all associated conditions, and form response strategies. More specifically, in facing an object as the incoming stimulus, we immediately exercise a mental process that consists of five sequential stages of operations:

1. perception of the incoming stimulus (thus focusing on the specific people or situation);
2. interpretation of the stimulus in reference to some previously learned *implicit semantic features* (thus assigning meanings to the stimulus);
3. integration of the meanings of the stimulus within the repertoire of all implicit semantic features;

4. formation of behavioural dispositions (or intents); and
5. expression of dispositions in terms of overt behaviours.

This five-stage operational mechanism generally applies to all incoming stimuli from both the subjective and objective cultures across a variety of situations, contexts, and times. Similarly, the manifestation of any new economic and social behaviour will involve the mental process of the assignment and integration of meanings and the subsequent determination of overt behaviours. This assignment–integration–determination process is defined as the psychosemantic process of an individual.

*Two evaluation systems*. The formation of any business-related behavioural dispositions has to premise on two evaluation systems: (1) the *cognitive evaluation* in terms of some commonly agreed upon descriptive semantic features and (2) the *affective evaluation* in terms of an individual's attitudinal and emotional features. For example, an international business in China will be cognitively evaluated as a profit-driven endeavour. But it may be affectively evaluated as *good* by some individuals, and as *bad* by others.

Within each society, descriptive criteria used for the cognitive evaluation exist practically for all social and economic behaviours (e.g., the production level as an index of successful manufacturing). Such cognitive features are usually stable over time and commonly used by the members of a society. They represent certain group-traits or stereotypes that may be favourable or detrimental to a new international business endeavour (Piontkowski *et al.*, 2000). Similarly, affective features may also exist, but will likely vary in applications, depending on individual differences in life history (e.g., the same product may be judged as equally likable and dislikable by different members of a social group).

In combination, the features from these two evaluation systems constitute an individual's repertoire of meanings, and thus determine his psycho-semantic process involved in international business-related behaviours. These features will enable the individual to interpret, acquire, or modify the meanings of the work environment, and thus formulate certain response intents and overt behavioural patterns. (Under this premise, the manager of a workplace can develop, on the basis of the two evaluation systems of all new employees, various action-oriented training programmes.)

Osgood and associates conducted extensive semantic differential research in 30 language/culture communities around the world (Osgood, May and Miron, 1975; Osgood and Tzeng, 1990). Humans' affective system has been found to comprise three universal semantic features. (1) *Evaluation* represents attitude toward an object (incoming stimulus or situation) as being good or bad, and nice or awful. (2) *Potency* indicates the felt intensity (strength) toward the object as being strong or weak, and potent or impotent. (3) *Activity* reflects the perceived energy coming out of the object as being active or passive, and fast or slow. These universal affective features are capable of predicting behavioural dispositions in various social and occupational contexts (Osgood and Tzeng, 1990).

Therefore, to succeed in any international business, the affective and cognitive evaluation systems of foreign delegates must be identified and compared with those of the co-workers in host Chinese culture.

*Five ecological environments.* Generally speaking, each international businessperson has to deal, simultaneously, with the characteristics of two nations – the foreign (hosting) and his indigenous (visiting) nation. The characteristics of each nation can be organized under five ecological systems:

1.  *Idiosystem of individuals* consists of the subject (such as personality traits) and objective characteristics (such as ethnicity and physical health) of individual citizens. The idiosystem defines a person's identity and serves as the foundation for developing other systems.
2.  *Microsystem of families* encompasses the characteristics of the memberships, values, functions, and conditions of individual households. These characteristics are significantly different between Western contemporary society and Eastern Chinese culture.
3.  *Exosystem of community or workplace* includes the characteristics of residential neighbourhood, occupational associations, and various intra- and inter-group activities. Specific subjective and objective characteristics of these conditions usually dictate the development of general life patterns for individuals and families in each workplace.
4.  *Macrosystem of nation* reflects salient characteristics of a nation (e.g., American individualism vs. Chinese collectivism). However, within-national differences also exist, especially under the recognition of societal diversities, such as ethnicity and political persuasions in the US. In conducting international businesses, all foreign workers are required not only to master the objective culture of the host nation (e.g., different banking routines) but also to address different subjective characteristics (e.g., state-right-centred legal foundations in China).
5.  *International geopoliticsystem* represents the roles, styles, dispositions, and behavioural patterns of all nations (or political entities) when they perform various international functions (e.g., in political, economic, and cultural affairs).

Each international businessperson brings these five ecological systems with him in a host nation. These systems are interrelated and simultaneously operating whcn he makes contact with members of the host nation. For example, a visiting American employee may be perceived by a Chinese co-worker as 'representative' of all Americans. Thus, although this American delegate may possess some unique characteristics at the personal and family levels (in idiosystem and microsystem), he may still be mistreated in terms of certain stereotypes of Chinese against all Americans (i.e., in his macrosystem).

*Ten culture-by-ecology units.* Based on the above separations of the two (subjective and objective) cultures and five ecologies, all living and working conditions of an American delegate in China can be organized into 10 $(2 \times 5)$ culture-ecology units for analyses. For example, an American employee may possess some salient characteristics in two culture-ecology units. First, in the subjective culture-idiosystem unit, this US delegate may strongly believe in open and direct communication, individuality, and self-actualization. Second, in the objective culture-macrosystem unit, this delegate may emphasize all kinds of material conveniences, such as spacious housing, modern utilities, and recreation facilities.

Generally speaking, the characteristics of an employee, while in his own nation, would predetermine his inclination, frequency, style, and satisfaction with an international occupation. Thus, the international company should have a comprehensive knowledge of all employees both before and after they physically station in a foreign facility. Such knowledge can be obtained in terms of (1) detailed identifications of all social-economic behavioural characteristics, and (2) a logical organization of these characteristics in terms of the 10 culture-ecology units.


## FUNDAMENTAL POSTULATES OF SOCIAL-ECONOMIC BEHAVIOURS

When applied to international businesses, the above theoretical consider-ations lead to the formation of *10 fundamental postulates* about the process usually involved in social-economic behaviours:

1. *Decomposition property.* The global economic behaviour in international contexts can always be decomposed into many organizational units (components) for assessment.
2. *Network capacity.* All components can be structured under some logical network that can specify as to what, when, how and why the individual components exist and interact.
3. *Continuing interaction.* The network of the decomposed components con-tinues to function over time and situations. However, its logical structure

   may change under the influence of continuing interactions with the host culture.

4. *Attribution principle*. Human behaviours can always be explained by some underlying attributes (semantic features). Such attributes are the primary targets of training for international businesses.

5. *Bipolarity and gradient*. Human attributes are bipolar in nature and continuous in gradients. Various descriptors commonly used in each language can reflect all attributes.

6. *Disposition determinant*. Social-economic behaviours are the manifestations of intentions (dispositions). Such intentions are based on the results of the decomposition, assessment, and integration of all available information.

7. *Positivism nature*. Human nature is predisposed to goodwill because each individual tends to pursue the fulfilment of the needs and motivations that are acceptable in the living environment.

8. *Evaluation of others*. In international interactions, people tend to evaluate the motives and behaviours of others in reference to certain projected social norms.

9. *Evaluative goals*. Evaluations of international behaviours are generally intended to explain the present, postdict the past, and/or predict the future circumstances. The ultimate goal is to direct and facilitate desirable economic behaviours.

10. *Reinforcement of consequences*. A person's overt behaviour always leads to certain perceived impact on others that will subsequently reinforce, positively or negatively, his own future psychosemantic process.


## PSYCHOSEMANTIC PROCESS MODEL

The above theoretical considerations are used to construct the Psychosemantic Process Model of International Social-Economic Behaviours (Figure 5.1). In this model, the general social-economic behaviours in international businesses are decomposed into *ten model components* that are integrated under six *functional stages*. As a whole, the model can be used to analyze any international behaviours in specific causal terms.

### Stage One: Background Characteristics

Each person, through a long-term social learning process in various ecological environments, develops some unique patterns of thinking, feeling, judging and behaving across a variety of situations. Such patterns usually

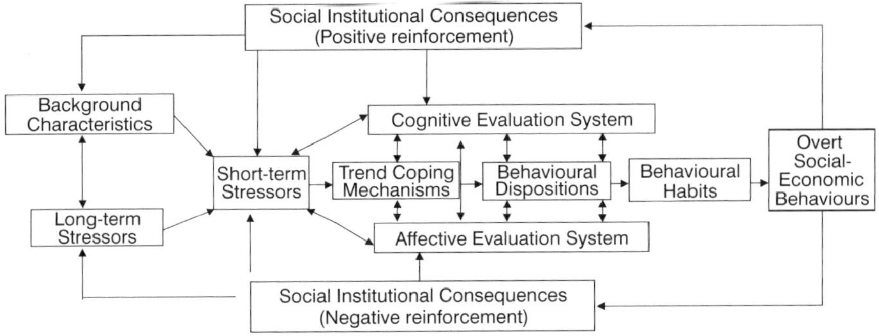


*Figure* 5.1 Psychosemantic process model of international social-economic behaviours

serve as foundations for engaging in international activities. These foundations are organized under two structure components of the Model:

- *Background characteristics component* comprises a person's 'normal' characteristics in the subjective and objective cultures over the five ecological units (the term *normal* means that the person's characteristics fall within the general trends of the society). Such normal characteristics enable the person to engage in international activities.
- *Long-term stressors component* means that a person has certain difficulties (usually deficiencies) in background characteristics that tend to have long-term and negative impact on international conducts (e.g., a physical disability or unfamiliarity with the language of a host country).

**Stage Two: Antecedent Stimuli or Circumstances**

This stage indicates the emergence of certain new events or conditions that require a person to respond immediately or eventually. The new events may vary due to individual differences, ranging from an acute sickness to a new job assignment in a foreign country. In general, such events will require new adaptations in feeling, thinking and behaving that may be different from a person's familiar background characteristics.

Such new emergencies (events or circumstances) are organized under the structure component of Short-term Stressors in the model. In international behaviours, the resolution of a short-term stressor will require a comprehensive analysis of other model components.

**Stage Three: Evaluation Processes**

In facing a short-term stressor, a person will generally go through the mental process of evaluating (1) the stressor and related circumstances and (2) his own capacities to adapt or change the stressor. In the model, two evaluation components are differentiated:

- *Cognitive evaluation* is to analyze the short-term stressor and related circumstances in terms of descriptive semantic features. Analytic results will be integrated and used to formulate expectations and response strategies; and
- *Affective evaluation* is to assess the short-term stressor, related circumstances, and response strategies in terms of a person's affective semantic features (such as the Evaluation, Potency and Activity dimensions identified in Osgood's cross-cultural research).

Both evaluations will interact and reciprocally reinforce in their conjoint contributions to the final selection of response strategies described below.

**Stage Four: Behaviour Generation and Monitoring Process**

In facing a short-term stressor and after completing the two necessary evaluations, the person will generally go through a new stage of generating and monitoring responsive behaviours. This stage comprises 3 operational sub-stages that are described as model components below:

- *Trend coping mechanism* represents the retrieval of some coping methods or strategies from life history in order to deal with the similar or new short-term stressor.
- *Implicit disposition* represents the generation of behavioural intent and motivation to respond with some new or existing coping methods or strategies.
- *Adjustment or monitoring of habit* represents the fact that the selected coping methods or strategies will be compared, adjusted or reinforced by a person's habitual acts under similar circumstances.

In essence, these sub-stages constitute a functional linkage across (1) the selection of relevant behavioural strategies, (2) the formulation of new behavioural dispositions, and (3) the adjustment of the intended acts in terms of prior behavioural habits.

## Stage Five: Overt Behaviour

This functional stage, as a model component, represents the expression of behavioural dispositions. All overt acts in international businesses have psychosocial implications when judged against the model components of the host culture.


## Stage Six: Social Consequences

All social and economic behaviours in the host country are interactive and have implications in all ecological environments (from personal to national levels). Thus, the overt behaviour of a foreign delegate will be judged against the characteristics identified in the 10 culture-ecology units of the host culture. Minimally, the behavioural consequences can be considered as socially acceptable or unacceptable, or economically normal or deviant. In either the positive or negative direction, the resulting consequences will have a reinforcement effect on the subsequent international behaviours. Therefore, such social judgmental process constitutes an important model component.

From the standpoint of the host culture, social consequences play an important role for the maintenance and new development of international businesses. Such a role is especially critical when two nations differ significantly in various ecological environments. For example, China and the US differ in political ideologies, cultural orientations, and economic systems. Evaluations of the same overt behaviour might generate incongruent or even opposing social consequences. Thus, businesspersons may unknowingly make detrimental decisions, if based solely on their own indigenous cultural standards.

In summary, the structural components of this model deal with issues involved in all 10 culture-ecology units at the same time, but to varying degrees. Although their structural characteristics are independent of each other, their functions are intimately related. Thus, when the model is used to address a specific international business, the structural characteristics of different components must be delineated in detail and integrated in terms of their specific functions.

Under this premise, this model can serve as an effective instrument to summarize all existing knowledge of social-economic behaviours in a foreign county. The resulting materials can further be used to formulate strategies to achieve desirable outcomes of certain economic behaviours.

## APPLICATIONS

In the contemporary one-world community, international business, tourism, and communications have greatly reduced the geographic boundaries and international distances. However, international differences remain in diverse objective environments, ranging from natural resources to economic systems. More important are differences in subjective cultural characteristics of philosophy, work ethics, and values. Such objective and subjective cultural differences are critical to the development and maintenance of any international businesses. All international entities need to have in-depth knowledge and skills in dealing with important similarities and differences with other nations.

### Three Types of Cultures: A Dyadic Perspective

When two nationals (e.g., a Western manager and a Chinese employee) form a working relationship in China, the separate psychosemantic processes of the two individuals will interact and generally result in three distinct types of cultures:

*Two separate, indigenous cultures.* The two individuals will uphold their indigenous national characteristics that may be organized in terms of the $2 \times 5$ (culture $\times$ ecology) units. In interactions, each individual will have to deal with two separate, indigenous cultures: one is the *first culture* (from his own national perspective), and the other is the *second culture* (from the counterpart's national perspective). These two cultures will provide foundations for the formation and maintenance of all subsequent social and economic relations.

*A conjoint culture.* Through working relations, each individual will compare, consciously or unconsciously, his own (first) culture against the co-worker's (second) culture. Similarities and differences will be identified across all five ecological systems. The comparison results may further be differentiated in reference to a bipolar (similarity–dissimilarity) continuum – ranging from *identical* on the similar pole to *opposite* on the dissimilar pole, with *incomparable* at the midpoint (neutrality) of the continuum.

When two indigenous cultures are similar, the co-worker will likely become a positive reinforcement in subsequent interactions. Otherwise, dissimilarities would require both individuals to understand, accept, adjust to, or ignore the discrepancies.

The process of reinforcing similar, as well as adjusting dissimilar characteristics is, in fact, to develop a *conjoint (third) culture* in each business organization. When the business association continues, the contents of the third (conjoint) culture will generally enlarge to harmonize some unique aspects of the two indigenous cultures. Such expansions will facilitate, and thus improve, the quantity and quality of social and economic interactions between the two working nationals. Generally speaking, the harmonization of two indigenous cultures is a continuous *acculturation process* undergone by all nationals in the same business organization (Searle and Ward, 1990).

## Between-Cultural Differences and Impacts on Ideal Business Behaviours

Similarities and differences in various cultural characteristics would determine the entire process of performing all economic activities. The impact ranges from the establishment and maintenance of operation facilities to the production and distribution of goods. More specifically, significant differences have been found in employment expectations, product standards, and conflict resolution styles between the Chinese and Western societies. The abilities in handling such differences by and large determine the survival of Western investments in China.

### Critical Traits in Business Operations

For any Western companies to better prepare for successful business operations in China, the critical tasks would require:

1. A comprehensive analysis of both indigenous cultures.
2. The identification of a *real* third (conjoint) culture based on similar characteristics of the two indigenous cultures.
3. The formation of an *ideal* conjoint culture for the business establishment that will set the goals for all workers to achieve their maximal productivities and satisfactions.
4. Development of organizational programmes and management strategies in order to reduce the gaps between two (real and ideal) conjoint cultures.
5. Assisting all employees to harmonize the desired ideal conjoint culture with their unique indigenous cultures (e.g. Chinese workers harmonize between the Western productivity-oriented company environment and the Eastern people-oriented working atmosphere).
6. Establishment of a programme to continuously monitor and improve the above tasks.

Note that different business entities may require different sets of characteristics for forming their *ideal* conjoint cultures. For example, in the

Chinese markets, an auto-parts production plant and the life-insurance company may require quite different qualifications of their employees in skills, aptitude and occupational interests. Therefore, for each programme development, such industry-specific characteristics should be identified in both objective and subjective cultures across the five ecological levels.


## Critical Comparisons between Societies

In the following, the Chinese and US societies are compared in terms of two important areas of job-related characteristics: general social/cultural orientations and family conceptions.

*Eastern vs. Western orientations.* The Eastern and Western (represented by the US) societies are different in historical backgrounds and cultural values (Triandis, 1990). Such differences are best summarized under the bi-polarities between individualist and collective social orientations and family laws shown in Tables 5.1 and 5.2.

The American culture is idiosyncratic or individual-centred, promoting individual achievement, self-reliance and independence. Moral standards and values are personal matters. Hard work is important only to the realization of personal aspiration (Schorr, 1976). Such idiosyncrasies are rarely committed to the community or even family. When an individual's free choice and satisfaction are threatened, the idiosyncratic individual tends to withdraw from exiting associations to find new alignments (Ellis, 1978). The *emic* concept of individualism is the ability to make free choice.

The traditional Chinese culture is gregarious and group-centred. Social-economic behaviours of an individual are for the ultimate pursuit of institutional prosperity. Individuals are mostly allocentric or situation-centred with the emphasis of mutual dependence and 'bondage' producing 'collective achievement'.

*Concepts of family.* The American culture stresses the importance of individuality as the pivotal unit of societal institutions. Thus, families exist for the pursuit and protection of individual growth and development. The traditional Chinese families are collectivist in nature and highly patriarchal. The parents, especially fathers, are the authority figures. Obedience to the elders is the most dominant and valued behavioural norm. Usually, the major goal of personal development is to attain resources for family survival (Ellis, 1978).

*Convergence between Eastern and Western cultures.* In contemporary society, distinctions between individualism and collectivism are not totally static (Triandis, 1988). In fact, due to rapid industrial modernization, media communications, and close international business activities, these two

*Table* 5.1   Comparisons on family values between traditional Chinese and contemporary American cultures

| Comparison units | Traditional Chinese | Contemporary American |
| --- | --- | --- |
| **A. CULTURAL/SOCIETAL POLARITIES:** | | |
| Societal orientation | Collectivism<br>Hierarchical status quo | Individualism<br>Equality |
| Organization unit | Family/gregariousness | Individual/idiosyncrasy |
| **B. SOCIAL-ECONOMIC BEHAVIOURAL POLARITIES:** | | |
| Behavioural principles | Li-Particularism<br>Allocentric | Moral ethics<br>Idiocentric |
| Behavioural goals and values | Collective/patriarchal<br>Traditional consistency<br>Bondage<br>Family survival<br>Conformity<br>Interpersonal harmony<br>for group maintenance | Individual realization<br>Future possibility<br>Hard work<br>Personal aspiration<br>Competition<br>Intrapersonal<br>satisfaction with<br>advancement |
| Behavioural characteristics | Personal/relative<br>(different behaviours<br>toward different classes<br>of people)<br>Situational adaptability | Impersonal/absolute (same<br>behaviour toward all,<br>predictable before<br>occurrence)<br>Situational invariation |
| **C. FAMILY CONCEPT POLARITIES:** | | |
| Formation of relations | Natural | Contractual |
| Concept of marriage | As childbearing partnership | As domestic relation |
| Law of marriage | Li-Chinese traditional customary law | Christian law |
| Obligation to family | Abstract concept | Concrete household |
| Family obligation | Whole family as entity | Other family members |
| Children's role | Filial duty, respect and obedience as obligations | Double dependent (on parents and state as right/entitlement) |
| Resolution of family dispute | Extrajudicial mediation (conciliation) (family council first) | Judicial determination (confrontation) |

92

*Table* 5.2   Cross-cultural comparisons of family laws

| Comparisons | Collectivist (Taiwan) | Revolutionary (Mainland China) | Individualist (USA) |
|---|---|---|---|
| **A. FAMILY LAW FOUNDATIONS** | | | |
| Cultural foundations | Confucian values (Hierarchical classes) | Equality of all individuals (no social class) | Western liberty (contractual theory) |
| Family law principles | Natural-Basic social/ human values | Marxism/Leninism | Institutional definition (nuclear family) |
| **B. SUBSTANTIVE FAMILY LAW DEFINITIONS** | | | |
| Marriage | Childbearing partnership | Domestic association | Contractual |
| Family composition | All household living members | Consanguinity natural by birth/ marriage | Contractual by marriage, parent-child living together |
| **C. LEGAL RIGHTS AND DUTIES** | | | |
| Function of protective laws | Defining family values (a priori for prevention) | General guidance for social order (prevention and intervention) | Defining social problems (after-fact sintervention) |
| Parent to minor child | Duties of support and property right | General protection as rights and duties | Parent's duty to provide |
| Support duties of adult child | Mutual (not reciprocal) Absolute duties of support (lineal ascendants) | Mutual and reciprocal | Reciprocal (not mutual) |
| Remedies for violation | Broad civil/ criminal<br>1. Direct negotiation<br>2. Family Council mediation<br>3. Civil suit<br>4. Criminal abandonment (enhancement, if abandoning parents) | General criminal abandonment | Specific civil/ criminal (also by court's contempt power) |

cultural orientations are converging toward forming a larger third culture. In this process, two indigenous cultures influence each other, and each faces unprecedented challenges from the other.

Thus, while working in China, an American delegate faces the reality that individualistic traits are confronted with various 'emergent' stressors of the traditional collective values. This American national in the Chinese workplace has to harmonize the typical individualist traits of hard work, success, individual achievement, competition, future orientation and absolute morality with the typical Eastern collectivist traits of conformity, sociability, sensitivity to others, and relativistic situation-centred norms. Similarly, the Chinese culture has lost its firmly bounded contexts in modern workplaces. The Chinese occupational environment is no longer a solid and well-defined unitary institution. Thus, in the same Chinese workplace, the American delegate cannot automatically assume that traditional values can be relied on, long-term membership can be established, and stable productivity can be expected.

*Other challenges in international behaviours.* Various social problems have emerged as common phenomena in many workplaces, e.g. prejudice, inter-group hostilities, and discrimination (Tzeng and Jackson, 1994). Such subjective cultural problems also become critical challenges in all contemporary international businesses. These problems have significant negative impact in two ways. First, each business association is forced to deal with two different indigenous cultures and simultaneously their continuing changes – either as social progresses or problems. Second, the positive convergence between the two indigenous cultures in the joint working environment will create many negative short- or long-term 'stressors' ('emergencies') that require all workers to adjust in other ecological environments. For example, Chinese workers may acquire Western individualist values in modern facilities. But in facing various contrary behavioural expectations from relatives and friends, the Chinese workers may consider the accustomed Western values as obstacles to the wholesome development of personality in the family and community.


## CONCLUSIONS AND RECOMMENDATIONS

For international workers, all economic behaviours are subject to the acculturation process of evaluation, adaptation and assimilation between two sets of national characteristics. The proposed model is to decompose such a process into many logically related units for detailed comparisons of any two indigenous cultures that may be defined in various ways. More specifically, for any Western company to send business delegates to a foreign

*Table* 5.3    Illustrative training contents

---

*A.  Prospective living situations of delegates and their spouses in a host country:*
1.  General social and environmental orientations – geographic environment, social systems and behavioural norms, housing, transportation, weather and seasons, commerce, food and eating patterns, general lifestyles, recreations, communications, and community involvement.
2.  Personal idiosystem – monetary behaviours, locus of control, religious orientations, management of time, occupational goals.
3.  Family microsystem – general family structure, kinship relations, gender and roles, children's roles, rights and reciprocal responsibilities.
4.  Conceptions in inter-personal relationships – private vs. public circumstances, in-group vs. out-group differentiation, intimacy vs. remoteness maintenance, associative/neutral/disassociative inter-group relations.

*B.  Social-economic behaviours of delegates:*
1.  General conceptions of occupation – affiliative vs. achievement orientation, success vs. failure standard, and supraordinate vs. subordinate relations.
2.  Economic-related conceptions – occupational prestige and commitment, commercial and economic productivity, formation of working habits, and problem-solving styles.
3.  Motivation and adaptation in international business environment – long-term stressors, affective and cognitive evaluations, copings, motivations, reinforcement processes.

---

county, the ideal preparation would be to conduct an intensive acculturation training for the delegates and their spouses.

The training should focus not only on work-related knowledge and skills, but also on the language and subjective characteristics of the host culture at various ecological levels. For illustration purposes, the contents of a training programme are outlined in Table 5.3.

To address these training topics, the model calls for multidisciplinary efforts to compare not only the visiting and host national backgrounds, but also their subjective cultural values and objective societal realities. For the ultimate success of an international business, this model encourages the business organization and its delegates to integrate all relevant information for maximal utilities.

In conclusion, the nature of international business is extremely dynamic. Any business endeavour will involve both the processes and the outcomes of acculturation of all employees. For this reason, the proposed model can become an invaluable guide to structurally establish a formative conjoint institution from two previously unrelated societies, and to functionally assess their converging relations. As the World Trade Organization (WTO) enlarges its membership, the covenants over all economic activities will become dominating forces to promote two simultaneous convergences: *first*, the within-national convergence among diverse (objective and subjective) cultural characteristics, and *second*, the between-national convergence between individualistic and collectivist societal orientations. For such convergences, the proposed model seems to offer useful measures for both the structural establishment and the functional monitoring of economic endeavours in a foreign country. Equally important is that the model calls for the harmonization between the *ideal* characteristics expected in the workplace and the *real* (idiosyncratic and traditional) characteristics existing in the living environments of family and community.

# References


Ellis, G. J. (1978) 'Supervision and Conformity: A Cross-Cultural Analysis of Parental Social Values', *American Journal of Sociology*, 84, pp. 386–403.

Hannigan, T. P. (1990) 'Traits, Attitudes, and Skills that Are Related to Intercultural Effectiveness and Their Implications for Cross-Cultural Training: A Review of the Literature', *International Journal of Intercultural Relations*, 14, pp. 89–111.

Oberg, K. (1960) 'Culture Shock: Adjustment to New Cultural Environments', *Practical Anthropology*, 7, pp. 177–82.

Osgood, C. E., May, W. H. and Miron, M. S. (1975) *Cross-Cultural Universals of Affective Meaning* (Urbana: University of Illinois Press).

Osgood, C. E. and Tzeng, O. C. S. (eds) (1990) *Language, Meaning and Culture* (Connecticut, Westport: Praeger).

Piontkowski, U., Florack, A., Hoelker, P. and Obdrzalek, P. (2000) 'Predicting Acculturation Attitudes of Dominant and Non-Dominant Groups', *International Journal of Intercultural Relations*, 24, pp. 1–26.

Schorr, A. L. (1976) 'Family Values and Real Life', *Social Casework*, pp. 398–411.

Searle, W. and Ward, C. (1990) 'The Prediction of Psychological and Sociocultural Adjustment during Cross-Cultural Transitions', *International Journal of Intercultural Relations*, 12, pp. 61–71.

Triandis, H. C. (1988) 'Cross-Cultural Training Across the Individualism-Collectivism Divide', *International Journal of Intercultural Relations*, 12, pp. 269–89.

Triandis, H. C. (1990) 'Cross-Cultural Studies of Individualism and Collectivism', in R. A. Dienstbier (ed.), *Nebraska Symposium on Motivation, 1989* (Lincoln: University of Nebraska Press).

Tzeng, O. C. S. (1993) 'The Psychosemantic Paradigm and its Love Models. A Theoretical Content Framework for Studies of Love', chapter 2 in Oliver C. S. Tzeng, *Measurement of Love and Intimate Relations* (Connecticut, Westport: Praeger).

Tzeng, O. C. S. and Jackson, J. W. (1994) 'Effects of Contact, Conflict, and Social Identity on Interethnic Group Hostilities', *International Journal of Intercultural Relations*, 18(2), pp. 259–76.

# 6 The Challenge of Confucius: The Generalizability of North American Career Assumptions

Cherlyn S. Granrose

## INTRODUCTION

According to much of the North American careers literature, in the twenty-first century people will have boundary-less careers moving among multiple organizations that are low in hierarchy, high in the use of technology, and flexible in responding to customer preferences in many parts of the globe (Arthur and Rousseau, 1996). In order to be competitive in the twenty-first century, organizations will have to confront the growing gap between rich and poor nations and learn how to prosper in the global marketplace by building organizational relationships with diverse stakeholders in ways that lead to mutual trust (Roberts, Kossek, and Ozeki, 1998; Zahra, 1999). During the twentieth century, career research focused predominantly on describing careers in North America and Europe. To meet the challenge of describing careers in the twenty-first century, career theory and research must take into account new organizational realities at all levels of analysis and must consider career experiences in all parts of the globe. This chapter uses excerpts from the writings of Confucius to identify assumptions about organizational life that permeate many East Asian organizations and describes some ways that these assumptions appear in contemporary Chinese organizations. It compares these assumptions with assumptions common in writings about North American careers and suggests possible characteristics for career theories that aspire to guide organizations doing business in North America and in East Asia in the twenty-first century.

Confucian philosophy was chosen because of the important role that Asian organizations arising from Confucian cultures will play in the new century (Woronoff, 1986; Abegglen, 1994). Despite the Asian economic crisis of 1997,

Asia continues to offer business opportunities that North American firms cannot ignore. The OECD predicts that China will overtake the United States as the world's largest economy (in GDP and population) and will be double the size of the economy of India (the world's next most populous country) by the year 2015. This means that China will regain the position it has held as the world's largest economy in nine of the ten centuries in the last millenium (Ning, 1998). In addition, Chinese expatriates have spread a bamboo network across the world. This network influences business in many Southeast Asian countries and has substantial influence in the economy of many parts of North America (Weidenbaum and Hughs, 1996). Thus the potential of Chinese influence on the global economy of the future is growing in importance.

In addition to Chinese influence created through the economic strength of ethnic Chinese businesses, the influence of Chinese cultural ideas extends to other Asian nations as well. China, or the middle kingdom (*Zhong Guo*), started to exchange cultural ideas and material artifacts with its neighbours before the Warring States Period (475–221 BCE) and trade was flourishing when Emperor Wu Di of the Western Han Dynasty adopted Confucian ideas as China's orthodox philosophy in 140 BCE (Shen, 1996; Ding, 1997). The philosophy of Confucius especially came to permeate the cultures of Japan and Korea as they struggled with China to create distinct identities and governments (Lee, 1984; Aoto, Fukuzawa, Hoyoshi and Yage, 1988). The combined economic influence of the People's Republic of China, The Republic of China, Japan, Korea, and Chinese expatriates in other countries is so large that people from cultures strongly influenced by Confucius make up one of the most potent forces in global business at the turn of this century.

A shared Confucian philosophical heritage cannot explain the prevalence or absence of economic growth in these nations with such different political and industrial systems (Biggart and Hamilton, 1997). However, when we speak of careers, we are analyzing not patterns of financial transactions, but human belief systems and structures and processes of human relationships that are profoundly influenced by the assumptions about human nature and human relationships outlined in Confucian precepts. Any career theory for the twenty-first century would be wise to address the beliefs of this philo-sophical system if it wants to be relevant to a substantial proportion of global business.

In order to identify the patterns of Confucius that have been retained in contemporary Chinese organizations, first we must describe briefly the human resources practices of two organizations on the opposite ends of the market economy versus the planned economy spectrum. We then can compare these practices with the philosophical principles of Confucius and with the career practices of Western organizations.

## HUMAN RESOURCE MANAGEMENT IN CONTEMPORARY PR CHINA

There are four general types of organizations in China: state-owned enterprises (SOEs), collectively owned enterprises (COEs), individually owned enterprises (IOEs), and other types of owned enterprises (OOEs) (Bell, *et al.*, 1993; Guo, 1999). Enterprises under the control of some governmental unit are considered part of the state sector (SOEs and COEs), whereas all others are considered part of the non-state sector, but these divisions are becoming increasingly blurred. Although all organizations are covered by central government laws and policies concerning how to manage their human resources, and thus some aspects of the HR policy descriptions apply to all organizations, in fact every organization implements these policies in a unique manner. This variation is due to different policies applying to different types of organization, to a lack of strict enforcement of government regulations, to a gradual and uneven implementation of market economy reforms, and to variation in the local labour force environment and history. The decentralization of management authority undertaken as part of economic reform also permits senior executives some latitude in how to interpret and implement government guidelines.

Interviews identifying common HR practices were conducted with human resource managers of 19 organizations in all parts of the PR China and all industry sectors. Responses from two organizations were selected to be presented here because they exemplify the range of contemporary human resources practices in China. The first is a traditional SOE manufacturing firm that has made marginal adjustments to the new market economy. The second is a private organization that is highly oriented to market economy principles. From these examples we can explore the extent to which Confucian assumptions prevail in Chinese organizations today.

### A State-Owned Tool-Manufacturing Organization

The SOE manufacturing organization was established in the 1950s as one of the major suppliers of heavy machine tools in the central planned economy. It is located in a provincial capital in central China and has a workforce of about 10,000 employees, 10 per cent of them women. It manufactures very heavy-duty machine tools such as boring tools, lathes, milling machines, turntables, and grinders.

Currently this firm supplies machinery to all parts of China and to more than 20 foreign countries but it has not located markets to absorb all of its production capacity. Like many SOEs, the firm faces serious, ongoing problems in making the transition to a market economy. The tangible assets (Y400 million or US$50 million) have lost their value because the technology has become outdated and machines and physical facilities have become worn. A limited number of manufacturing lines have been updated using nine imported computer work-stations to implement computer-aided design, computerized manufacturing processes and computerized quality control. Other outdated facilities have been retired from use and the workers redistributed. One-third of the workforce retain their positions in running the business; one-third provide life support services to other members of the work unit by running the hospital, store, restaurants, schools, and other support services supplied by the SOE to its workers; and about one-third are excess workers that have been distributed into various categories of less than full employment. The local personnel board will not permit permanent layoffs, much to the distress of the firm's president, who is charged with making his firm profitable.

The central government entity responsible for this firm is the Ministry of Machinery and Industry. At present the factory is still owned by the central government but the responsibility for government aspects of management has been decentralized to the municipal branch of this ministry. The firm's president reports to the municipal Bureau of Machinery and Industry, and the firm's Communist Party chairman reports to the municipal head of the Communist Party. Information about human resources in this organization was provided by the president of the firm and corroborated by an employee of the human resources department.

*Human Resource and Career Practices*

The Municipal Personnel Board does all strategic human resources planning for the number of people assigned to the firm but the internal personnel department does placement of personnel according to the HR plan submitted by each factory sub-unit. The firm is not recruiting any new workers at this time but it does receive an annual allocation of 30 people per year assigned to them as cadres from the Municipal Labour Bureau. Most of these are ex-military workers or new university graduates who have not found jobs on their own. Selection is based on the criteria of male gender, recommendations of past superiors, examination of the official personnel folder, as well as having skills in the specific field desired by the firm. There is one month of general orientation training given by the training department and the head of the department where the new worker is assigned. Each worker is given more on-the-job training by his or her supervisor. Little other formal training occurs.

Employees are motivated to work by receiving bonuses for high performance and by the use of moral encouragement via many meetings and factory posters. A typical poster seen on a factory tour included the following:

'Give up bad habits that destroy the reputation of the enterprise!'

'Follow Wu Tian Xiang (a model worker): Love your own occupation, struggle against difficulties, contribute your energy selflessly.'

The performance of factory workers is evaluated every month based on productivity quantity and quality. Performance of managers is evaluated based on self-appraisal, subordinate appraisal, and supervisor appraisal of personal potential (*nen*), numerical job performance (*ji*), and attitude and general behaviour on the job (*xin*). The party chairman evaluates moral character (*de*). Compensation and fringe benefits are given according to national base rate requirements plus some performance bonuses. Promotions are rare in the situation of there being too many employees, but if a promotion is given, it is based upon education level, CCP membership, and *guanxi* or personal relationships and exchange of favours.

**A Chinese Biotechnology Process High-Tech Conglomerate**

The Biotechnology High Tech Group is a nine-year-old private enterprise conglomerate of biotechnology, electronics, and real estate firms with over 2,000 employees, including 515 permanent employees and 1,500 contract workers. Ninety per cent of the headquarters employees possess a college degree. This firm produces biological and electronics products for sale in PR China and abroad. Stock of this firm began to be traded on the Shanghai stock exchange in the spring of 1997 when 375 million shares were offered on the stock market with an initial value of ¥1. According to the stock-offering prospectus, the total assets of the firm are worth ¥1.4 billion. Information about human resource policies was obtained by interviewing the senior HR manager and other members of the HR department.

*Human Resource Practices*

New employees are recruited through employee referral, direct application, college recruitment, advertisements in newspapers and TV, and the local Talent Market (a job fair run by the Municipal Personnel Board). A small group of managers called the examination council conducts the hiring interviews and makes the final hiring decisions. Individuals are hired if their background as stated in their personnel folder, interview and examination scores reveals sound personal characteristics, a high level of education and experience relevant for the position being filled. Some people believe personal relationships and favouritism play a role in this hiring procedure, despite the use of examination and committee hiring procedures.

Little formal training is provided, since employees are expected to have obtained a college education and relevant job skills at the time of hiring. Occasionally some short-term classes and training seminars are held, but most of the training is on-the-job and through the aid of a mentor or coach.

Performance appraisal occurs privately every month but employees are not always told the results. Performance bonuses vary from a letter of praise or small gifts for important holidays to substantial bonuses for spectacular performance. Employees are paid higher than average salaries and are expected to purchase benefits from this salary. It is possible to be fired from this company, although it is done only rarely if an employee survives the three-month probationary period. The career paths are varied with some senior managers expecting to stay in their current positions indefinitely, some younger employees getting promoted in less than a year, while others are hoping for promotions about every 4–5 years. A promotion is obtained through personal recommendation by a supervisor to the general manager, with the executive director's office committee having final approval. The Board of Directors approves promotions of top management. The formal criteria for promotion include workmanship, capability, diligence, good character, and achievement, as well as good relationships with co-workers and superiors.

As we will see, the characteristics used to evaluate personal potential and career mobility in contemporary Chinese organizations are remarkably consistent with Confucian teachings from 2,500 years ago. These descriptions of Chinese organizational practices will serve as examples to give substance to the following exploration of Confucius.


## CONFUCIUS AND THE TEACHINGS

From 140 BCE until the founding of the People's Republic of China in 1949, Confucianism has been the official mainstay of Chinese philosophical thought. Kong Qiu, Zhong Ni, Kong Fu Zhi, or Confucius, as he has been called at various times, lived in the kingdom of Lu in the middle Yellow River valley (what is now Shandong province in south-eastern China) from 549 BCE to 479 BCE, during the Spring and Autumn dynastic period. The impoverished son of a warrior who died when Qiu was three, Confucius was minimally trained in the traditional gentlemanly arts of music, archery, mathematics, calligraphy, ceremonial rites, and horsemanship. He probably received some education in the classical writings of the time: *The Book of Songs*, *The Book of History*, *The Book of Rites* (customs, norms and rules of behaviour), *The Book of Changes* (fortune-telling anagrams today often equated with Taoism), *The Book of Music*, and *The Spring and Autumn Annals* (historical and court records of the period). He performed military service under Duke Jiao Gung, and travelled in this capacity from about the age of 30. At some time during this period he began to teach. His circle included about 20 faithful disciples with as many as 70 or more other students finally attributed to his mentorship.

By the time of Confucius, there was great turmoil in the political life of China as leaders of the five powerful regions, Qi, Song, Jin, Qin, and Chu, struggled for dominance. Power was equated with control of land, and a single ruler did not dominate the country. Peasants were hard-pressed by demands for a large portion of their agricultural produce as the noblemen rulers required more and more resources to carry out these great struggles (Jian, Shao and Hu, 1986). The political struggles were mimicked by philosophical disputes. The philosophical struggles of the 'one hundred schools of thought' included the students of Confucius – Men Zi (372–289? BCE) and Xun Zi (310–238? BCE) – who sought to spread Confucianism; the Taoists – Lao Zi (365–290? BCE) – who advocated a return to a natural way of living apart from governmental rule; and the Legalists – Shang Yang (390 to 338? BCE) and Han Fei (280 to 233? BCE) – who advocated a centralized monarchy and the rule of law. This ideological struggle was influential in shaping the under-standing of Confucius we have today. It was the legalists who were responsible for the system of formal examinations for court service in later dynasties and who institutionalized mastery of Confucian teachings as part of the learning necessary to attain high status. The Taoists perpetuated beliefs in harmony even as they opposed the Confucian view of the importance of government. The followers of Confucius were continually responding to these alternative points of view as Confucian teachings were handed down.

Confucius did not create any original writings that have survived. His students and later followers compiled *The Analects* as a set of dialogues, sayings, questions and answers that have been much revised and interpreted in later years. Most of the original work was lost in the burning of the imperial library at the fall of the Qin dynasty. Thus, what we have today represents the reconstructed collected wisdom of Confucius, his students and later editorial scholars influenced by the philosophical and political struggles of the centuries since his death.

In the time of Confucius, people believed that the emperor as the personification of the 'Will of Heaven', their ancestors, who could return from death to guide or haunt descendents, and spirits of the natural world, influenced and controlled human destiny. In contrast to this view, Confucius advocated the power of moral thought and behaviour as a major force defining what it means to be a human being. His position was secular, and he was critical of those who later in his life sought to elevate his teachings to the level of spiritual guidance. He insisted that his goal was to give guidance to people about how to live their daily lives well and how to encourage good terrestrial government without influence from the celestial or spiritual world.

Confucian writings emphasize four areas: the supreme importance of living a moral life, learning and self-development, creating proper, harmonious relationships, and the duties of a just government. From these broad areas, in the following sections I have selected quotes that address issues important to careers theory. Since references to Confucius in management literature rarely provide original quotes, and since there are so many conflicting interpretations of each passage, I have included multiple passages for each topic. Quotes and the identifying book and chapter references (that is, 1:2) are taken from Ames and Rosemont, *The Analects of Confucius: A Philosophical Translation* (1998). (See author's note at the end of the chapter for full information about sources and translations.)

## Characteristics of a Virtuous Person

Essential principles of humanity or characteristics of the true 'gentleman' or 'good person' are empathy and loyalty as well as diligence, tolerance, and deference or humility. Other characteristics of a virtuous person include generosity, propriety, wisdom and trustworthiness. The following quotes illustrate Confucius' views on human nature and show how the emphasis of the text is about living a virtuous life or following The Way. The Way or Dao might be defined broadly as the life career, not confined to the work domain.

The Master remarked that Zichan accorded with The Way of the exemplary person in four respects: He was gracious in deporting himself he was deferential in serving his superiors, he was generous in attending to the needs of the common people, and he was appropriate in employing their services. (5:16)

The Master said 'Zeng, my friend! My Way is bound together with one continuous strand.' Master Zeng replied 'Indeed.' When the master had left, the disciples asked, 'What was he referring to?' Master Zeng said 'The Way of the master is doing one's utmost (being true to our nature, being loyal) and putting oneself in the other's place (benevolence, empathy), nothing more.' (4:15)

'A person who is able to carry into practice five attitudes in the world can be considered authoritative (have perfect virtue).' 'What are these five attitudes?' asked Zizhang. Confucius replied, 'Deference (respect), tolerance (magnanimity), making good on one's word (sincerity, fidelity, trustworthiness,) diligence and generosity (kindness). If you are deferential, you will not suffer insult; if tolerant, you will win over the many; if you make good on your word others will rely upon you; if diligent, you will get results; if generous, you will have the status to employ others effectively.' (17:6)

These characteristics of a virtuous person have been quite persistent and appear in many organizations as common criteria for hiring, promotion and honour in contemporary Chinese organizations.

**Enacting the Role of a Virtuous Person**

Ritual Propriety covered even minute details of dress, facial expression and speech. This referred not only to exact rituals of court or religion but also to the general way one conducted oneself in all circumstances. In this sense, the virtuous life was seen as a performance, and adherence to the norms for actions enabled one to live in harmony with others and with The Way.

'Do not look at anything that violates the observance of ritual propriety; do not listen to anything that violates the observance of ritual propriety; do not speak about anything that violates the observance of ritual propriety, do not do anything that violates the observance of ritual propriety.' (12:1)

Master Zheng said: '...There are three things that exemplary persons consider of utmost importance in making their way: by maintaining a dignified demeanour, they keep violent and rancorous conduct (cruelty and arrogance) at a distance; by maintaining a proper countenance, they keep trust and confidence near at hand; by taking care in choice of language and mode of expression they keep vulgarity and impropriety at a distance.' (8:4)

This particular aspect of a virtuous person, namely appropriate public behaviour as well as restraint in demeanour and language are aspects of moral character recorded in the contemporary PRC personnel record that follows each Chinese person from birth to death. This document is consulted before key career moves for any employee and a bad record will prevent access to good positions.

## Life Goals

In Confucian thought, the primary life goal is the cultivation of character as one follows The Way, not so that one can accumulate wealth or power. But wealth as appropriately earned in a well-governed state may be acceptable and *The Analects* states that by striving to follow The Way, poverty may be avoided as a matter of course. This last belief is in contrast to the state of poverty that seemed to follow Confucius during several parts of his life.

> The Master said, 'Wealth and honour are what people want, but if they are the consequence of deviating from The Way, I would have no part in them. Poverty and disgrace are what people deplore, but if they are the consequence of staying on The Way, I would not avoid them.' (4:5) (See also 7:12 and 14:12)

> 'Exemplary persons understand what is appropriate; petty persons understand what is of personal advantage.' (4:16)

> The Master said, 'To eat coarse food, drink plain water, and pillow oneself on a bent arm – there is pleasure to be found in these things. But wealth and position gained through inappropriate means – these are to me like floating clouds (uncertain and insubstantial).' (7:16)

> The Master said 'Exemplary persons make their plans around The Way and not around their sustenance. Tilling the land often leads to hunger as a matter of course; studying often leads to an official salary as a matter of course. Exemplary persons are anxious about The Way, and not about poverty.' (15:32)

The advice to remain on the path to virtue in the face of temptations to pursue wealth is as challenging today as it was in the time of Confucius. Many individuals see the transition to a market economy in China as a perfect opportunity to pursue wealth, but it is interesting to note that often individuals use some of this newfound wealth in ways that indicate that they still remain interested in Confucian virtues, like providing a special education for their children.

## Humility in Attaining a Position and a Reputation

Confucius sees gaining a position as something that happens when one is well prepared. Thus one should not worry about gaining a good reputation so others will know how well you can do, a violation of the value of humility, but one should concentrate on preparing well so that when a position is attained, one can fill it well. Reputation is based on actions, not words.

> The Master said, 'Do not worry over not having an official position; worry about what it takes to have one. Do not worry that no one acknowledges you; seek to do what will earn you acknowledgement.' (4:14) (See also 14:30 and 15:19)

> The Master recommended that Qidaio Kai seek office. He replied, 'I am not sure I am adequate to it.' The Master was pleased. (5:6)

> Zizhang inquired, 'What does the scholar-apprentice have to do to be described as being "prominent"? 'What can you possibly mean by being "prominent"?' replied the Master. 'One who is sure to be known, whether serving in public office or in the house of a ruling family,' answered Zizhang. 'That is being known,' said the Master, 'it is not being "prominent". Those who are prominent are true in their basic disposition, and seek after what is most appropriate. They examine what is said, are keen observers of demeanor, and are thoughtful in deferring to others. They are sure to be prominent whether serving in public office or in the house of a ruling family. As for being merely known, they put on appearances to win a reputation for being authoritative (virtuous) while their conduct belies it. They are wholly confident that they are authoritative, and sure to be known, whether serving in public office or in the house of a ruling family.' (12:20)

> Zigong asked about exemplary persons. The Master replied, 'They first accomplish what they are going to say, and only then say it.' (2:13) (See also 14:27).

> The Master said, 'Exemplary persons despise the thought of ending their days without having established a name.' (15:20)

Maintaining a good reputation is essential for good career opportunities in PR China. The use of the personnel folder as an official record of reputation institutionalizes the persistence of this cultural value. One notable remnant of the Cultural Revolution that is not consistent with Confucianism is the practice of recording not only deeds but also words as an indicator of a good reputation.

**Learning Skills and Character**

Learning is often characterized as a central Confucian tenet but it is clear from these passages that it is not formal education or learning specific skills that is important. The basic abilities are a given. Confucius advocated life-long learning from others and from introspection as a process of seeking The Way. If a person conducts himself or herself properly, informal knowledge will also become apparent. Later passages claim that without learning, one cannot really be a virtuous person even if one strives to have virtuous qualities.

> The Master said, 'To quietly persevere in storing up what is learned, to continue studying without respite, to instruct others without growing weary – is this not me?' (7:2)

> The Master said, 'In strolling in the company of just two other persons, I am bound to find a teacher. Identifying their strengths, I follow them, and identifying their weaknesses, I reform myself accordingly.' (7:22)

> Master Zheng said 'Scholar-apprentices cannot but be strong and resolved, for they bear a heavy charge and their Way is long. Where they take authoritative conduct as their charge, is it not a heavy one? And where (since) their Way ends only in death, is it not indeed long?' (8:7) (This is similar to ideas expressed in 6:27)

> The Grand Master asked Zigong, 'Your master is a sage, is he not? Then how is he skilled in so many things?' Zigong replied, '*Tian* definitely set him on course to become a sage, but he also has many skills.' On hearing of this, the Master said, 'The Grand Minister certainly knows me! We were poor when I was young, so I learned many a menial skill. Does an exemplary person have these skills? I think not! (He does not need them).' (9:6) (See also 9:2 and 19:7)

> Ziqin asked Zigong: 'When the Master arrives in a particular state and needs to learn how it is being governed, does he seek out this information or is it offered to him?' Zigong replied: 'The Master gets all he needs by being cordial, proper, deferential, frugal (restrained), and unassuming. Perhaps this way of seeking information is somewhat different from how others go about it.' (1:10)

The concept of continual learning from life experiences is institutionalized in all PR Chinese organizations, although the extent to which the organization provides formal training to supplement this is quite varied. The prevalence of the high value of education is demonstrated by the reliance on formal educational achievement as a hiring criterion for new managers, even though many older managers were not able to attain this level of education.

## The Process of Developing into a Virtuous Person

Passage through the life span is defined in terms of development of good conduct, but hope seems lost if one has not learned how to behave before the age of 40. By diligent study to develop character, by learning to follow ritual propriety, and by striving for empathetic helpfulness rather than lust, conflict, greed and status, one could become virtuous. This process is expressed in the form of meeting different challenges as one ages. Deference to elders is tempered by respect for the young and high standards of striving to follow The Way.

> The Master said: 'The young should be held in high esteem. After all, how do we know that those yet to come will not surpass our contemporaries? It is only when one reaches forty or fifty years of age and yet has done nothing of note that we should withhold our esteem.' (9:23) (See also 17:26 for similar sentiments)

> The Master said, 'From fifteen my heart-and-mind was set upon learning; from thirty I took my stance (was established); from forty I was no longer doubtful; from fifty I realized the propensities of *Tian* [understood the will of heaven]; from sixty my ear was attuned (to the reception of truth, to obedience); from seventy I could give my heart-and-mind free rein without overstepping the boundaries.' (2:4)

> Confucius said 'Exemplary persons have three kinds of conduct that they guard against: when young and vigorous, they guard against licentiousness (lust); in their prime when their vigor is at its height, they guard against conflict (temper); in their old age when vigour is declining, they guard against acquisitiveness.' (16:7)

> ... authoritative persons establish others in seeking to establish themselves and promote others in seeking to get there themselves. Correlating one's conduct with those near at hand can be said to be the method of becoming an authoritative person. (6:30) or (Now the man of perfect virtue, wishing to establish himself, seeks also to establish others; wishing to be enlarged himself, he seeks also to enlarge others. To be able to judge of others by what is nigh in ourselves – this may be called the art of virtue.)

This last passage has been interpreted as the Confucian version of the Golden Rule, but it was added more than 200 years after the life of Confucius but before the Christian era, probably about 262 BCE and may represent later controversies in the Confucian community of scholars. Acknowledging normal developmental processes as a way of growing into a virtuous life is consistent with promotion on the basis of tenure rather than performance. This personnel practice is being changed as organizations pay more attention to performance as a criterion for promotions in contemporary Chinese organizations. Recently, the government has made a strong effort to promote people under the age of 40 to leadership positions. This is a way of counteracting the educational deficits of older employees who did not obtain a good education during the Cultural Revolution but it also is consistent with the Confucian notion that it is possible to find virtuous young adults. The idea of promoting others as a way of establishing one's own virtue is an indicator of the importance of relationships and an early statement of contemporary practices of *guanxi*.

## Building Relationships

According to Confucian thought, one builds relationships only with those of good character and for pleasure, not for personal advantage or attaining a position. In developing relationships one must be humble, trustworthy, harmonious, and knowledgeable in order to have something of value to offer to your friends.

> The Master said, 'As a younger brother and son, be filial at home and deferential in the community; be cautious in what you say and then make good on your word; love the multitude broadly and (but) be intimate with those who are authoritative in their conduct (cultivate the friendship of the good). If in so behaving you still have energy left, use it to improve yourself through study.' (1:6) (See also 16:4 and 19:3)

> Master Zheng said, 'The exemplary person attracts friends through refinement (culture) and thereby promotes authoritative conduct (virtue).' (12:24)

> Confucius said, 'Finding enjoyment in three kinds of activities will be a source of personal improvement; finding enjoyment in three other kinds of activities will be a source of personal injury. Ones stands to be improved by the enjoyment found in attuning oneself to the rhythms of ritual propriety and music, by the enjoyment found in talking about what others do well, and by the enjoyment found in having a circle of many friends of superior character; one stands to be injured by finding enjoyment in being arrogant, by finding enjoyment in dissolute diversions, and by finding enjoyment in the easy life.' (16:5)

> The Master said: 'How can one possibly join with a common fellow in serving one's lord? Before such a person has been appointed to office, he is desperate that he will not be. Once he has been appointed, he is desperate that he will lose his position. And in his desperation that he might lose his position, he will stop at nothing to hang on to it.' (17:15)

The importance of having a good relationship with many good people has been expanded far beyond the original Confucian idea in contemporary China. Today the mutual obligation relationships of a well-developed *guanxi* network are part of the process of getting, keeping, and advancing in any job or of starting a new business of your own.

## Good Government

Good government is equated with the ruler acting like a moral person and not exploiting those under him. Each person should properly fill his or her role. The relationship between good government and living a moral life is that one should support the state when it is just and leave or remain silent when it is badly run. In the era of Confucius, the only formal organization was the governmental system, so his advice about how to be a good ruler also might be taken as advice about how to be a good manager today. The relationship between government and organizations is becoming more distinct in contemporary China but a strong sense that there is not a great separation between government and organizations persists even in private organizations.

> The Master said, 'Make an earnest commitment to the love of learning and be steadfast to the death in service to the efficacious Way. Do not enter a state in crisis and do not tarry in one that is in revolt. Be known when The Way prevails in the world but remain hidden away when it does not. It is a disgrace to remain poor and without rank when The Way prevails in the state; it is a disgrace to be wealthy and of noble rank when it does not.' (8:13)

> Duke Jing of Qi asked Confucius about governing effectively. Confucius replied, 'The ruler must rule, the minister, minister, the father, father, and the son, son.' (12:11)

> Zizhang asked about governing effectively, and the Master replied 'Be unflagging in deliberating about policy, and do your best in carrying it out.' (12:14) (See also 12:17, 12:19 and 13:1)

Zhonggong was serving as the steward in the House of Ji and asked about governing effectively. The Master said to him 'Set an example yourself for those in office, pardon minor offences, and promote those with superior character and ability.' 'How do you recognize those with superior character and ability in order to promote them?' Zhonggong asked. The Master replied, 'Promote those that you do recognize with the confidence that others will not spurn those that you do not.' (13:2)

This description of good government principles applies equally well to the contemporary practice of promoting your friends in organizational life in China.


## A SUMMARY OF CONFUCIAN PRINCIPLES AND ASSUMPTIONS IN HUMAN RESOURCES PRACTICES IN CHINA

The assumptions of Confucianism identified in the analysis of the writings that survive today have many parallels in the actual practices of contemporary organizations in China as illustrated in the two organizations discussed earlier in this chapter, and others familiar to the author. Although certainly there are clear differences in the market orientation of the high-technology firm and the government-owned machine-tools firm, some recognizable practices are consistent not only with the current Communist or Socialist ideology, but also are consistent with the teaching of Confucius. The major areas of consistency include the following: (1) Adherence to morality and virtue as key individual and governmental goals; (2) Attainment of higher position through study, maturity, recognition of superiors, and fulfilling loyal obligation relationships with co-workers and superiors; (3) Government primacy in defining roles and allocating people into specific positions in the social hierarchy; (4) Use of recognition and praise as common rewards and shame for not behaving in a virtuous manner as common punishments.

The company goals, rewarded individual characteristics and criteria for hiring and promotion include obtaining advanced education; exhibiting humility, selfless loyalty, and virtuous moral integrity; struggling against adversity toward perfection; and demonstrating appropriate behaviour (as well as particular skills); all key characteristics of the virtuous person in Confucian writings. In spite of the waning influence of the CCP in general, the party official, who has great power to damage individual lives by placing negative information in the personnel file that follows every worker for life, evaluates moral character. This serves to emphasize the relative importance of moral character over other valued characteristics. Although a person might argue that these virtues are always desirable characteristics in employees, regardless of the influence of Confucius, the pre-eminence of morality as the stated goal in the face of widespread corruption and lack of virtue among high organizational and public officials has eerie parallels to the situation during the Spring and Autumn period that Confucius addressed.

The primacy of relationships as the key means to promotion and recognition persists in spite of stated criteria of education and skill levels. Even in the market organization referred to earlier, the HR manager was keenly aware that the carefully developed skill tests and performance measurement practices could be circumvented by a person who had developed an obligation relationship with one of the firm directors. In the SOE the person with whom one needed to develop the relationship was different (the CCP representative in the HR department) but the principle remained the same. If you are known to be loyal, and in an obligation relationship with someone above you in the organization, your virtue will be recognized and rewarded.

The nature of the relationships today may differ from the ones described as ideal by Confucius. Confucius advocated developing relationships with virtuous people so one's own character, learning and reputation would be enhanced; and honouring existing relationships arising from family and one's roles in society and government; to promote social harmony. Today, relationships held according to one's position in the family or government certainly carry obligations of mutual loyalty and propriety, and following these obligations does promote social harmony, as advocated by Confucius. The relationships are hierarchical in nature so that each person is bound in a network of relationships defined by social position. The extent of obligation, and traditions that determine which obligations take precedence over others, are very strong and often misunderstood by an outsider such as an American business partner. However, additional relationships outside of those formed through family and work roles often are developed not only to increase one's virtue and reputation, but also to attain wealth and power, goals Confucius subordinated to moral development but lamented as common in his experience.

Use of praise to reward loyalty and good performance and shame to limit undesirable behaviour is built into Confucian doctrine and remains central to most contemporary Chinese organizations. True, financial bonuses have become added incentives in the organizations most strongly committed to the market economy so the firms can attract educated employees when the organization does not provide the same social welfare benefits offered by SOEs and COEs. The use of public positive and negative feedback, and motivation sessions to encourage delinquent employees to conform to organizational norms and obligations, is common in all types of organizations in China. Often this reliance on praise and blame is misunderstood by US organizations who do not realize that offering monetary rewards but no praise may not be as valued as they expect it to be based on experience with domestic-country employees.

The many discussions in Western literature of the importance of 'face' in Asian interactions are an outgrowth of the same value system. Confucius recognized that reputation for virtue is one's most valuable asset and rewards that increase this virtue are priceless above gold. Only those who cannot gain a positive reputation through virtue seek the secondary pleasures of wealth and power. Even the most powerful organizational and governmental officials, obviously well rewarded financially, recognize the cultural value of demonstrating personal goals of virtue, not wealth, and receiving praise for these virtues. They address this desire for praise as virtuous men indirectly, by demonstrating their reputation for having other gentlemanly virtues such as learning, skill in poetry and calligraphy via the elaborate scrolls that decorate their offices and public reception rooms.

The primacy of the government in assigning people to positions and in defining roles is one characteristic that is changing rapidly in contemporary China, but many remnants persist. Formal organizations certainly exist in addition to the government, but each role, expressed as a job classification and basic salary range, has a government definition, even in market and other non-governmental organizations. The government restricts the firing of people from their jobs, to prevent gross official unemployment, and must approve all transfers from one organization to another, even though the primary responsibility for finding a new job has shifted to the individual.

The role of the government as the final agency of the welfare of its citizens is supported by the Confucian doctrines as well as by socialist ideology so it is impossible to attribute this situation solely to Confucian heritage. But one could certainly argue that the presence of a Confucian heritage made these socialist principles more socially acceptable by the moral citizens at the time of the Communist revolution. Even under the strain of massive unemployment and rapid social change that accompany the present market transition, both the government officials, and more importantly the Chinese people, hold the government ultimately responsible for their welfare to an extent unknown in the United States.

The evidence for remnants of Confucian principles in contemporary Chinese organizations is consistent and positive but not exclusively causal. Certainly it is possible for alternative explanations to be used to justify each of these findings. Interestingly, these values persist in the face of many brutal campaigns to denounce Confucianism as a sign of past decadence leading to the corruption of the previous dynasty. In addition, some of the ways Confucian ideas are expressed in specific policies indicate areas of potential misunderstanding between US and Chinese businesses and others reveal possible areas of congruence.


## CONFUCIAN PRINCIPLES AND ASSUMPTIONS RELEVANT TO CAREERS

From these extensive quotes and the examples from contemporary Chinese organizations, several principles that express a specific code of values is evident. The major life goal is to strive to learn continuously how to live a virtuous life by following The Way. How much wealth, reputation or status one achieves is less important than following The Way with humility and propriety. What a person does is more important than what a person says, but once something is said, one should be sure to live up to these words.

Some assumptions that support these values have career implications. In completing the explanation of their philosophical translation, Ames and Rosemont (1998) have indicated some of the assumptions residing in the Chinese language and culture that also are helpful in clarifying career-relevant assumptions. They take as their basic tenet that the Chinese way of viewing the world is one of constantly changing events rather than stable things. Thus, things are defined in terms of changing relationships rather than enduring properties and The Way represents the *process* of becoming a virtuous person rather than the *achievement* of becoming one.

This way of looking at the world embraces the following assumptions:

1. The human being is not something we are; it is something we are becoming (see Ames and Rosemont, 1998, p. 49). In the process of becoming, human development occurs in stages with certain tasks and temptations common in each age.
2. The 'individual life in relationship' is the most important level of analysis. The basis for this assumption lies in the nature of a particular Chinese character as well as in particular passages of the text. 'Ren' is the character translated as 'the authoritative person or the virtuous person' in the quotes above and is composed of the character for one person and the character for two. Many of the relationships are hierarchical, such as father–son or ruler–citizen, but others are peer, as in friend–friend.
3. The important organizations are government and family. Note that Confucius has much to say about just government and the role of virtuous governmental leaders, but organizations other than government are

almost completely absent from *The Analects* because they were absent from the society at the time Confucius was teaching. The government only occasionally is discussed as an employer, and more often as an institution with obligations to the people and obligations of people to fulfil their proper roles in society. The other primary organizational unit mentioned repeatedly in *The Analects* is the family, and this group also is mentioned in terms of relationship obligations rather than occupations.

4. The nature of human relationships is one of mutual obligation and helpfulness. This relationship is conducted as an expression of integrity or self-realization, not one of subjugation. A person contracts obligations as part of being in every relationship.

5. Relationships are a result of the natural order, and not of personal striving. One fulfils relationship obligations in ritualized, proper, or normative ways to promote harmony in the relationships.

6. The function of learning is to facilitate the process of becoming an aware human being. Thus learning is an essential process of life rather than something done only in youth or only to obtain special skills.

Given these texts of Confucius and the assumptions that seem to underlie the teachings, as well as the extent to which these principles and assumptions emerge in contemporary Chinese organizations, the next question is the extent to which these teachings are consistent or inconsistent with contemporary career writing and practices in the United States.

## NORTH AMERICAN BOUNDARY-LESS CAREER THEORY AND RESEARCH

The large body of career writing in North American practitioner and academic literatures is replete with examples of common North American career principles and there is no one specific authoritative text, as there is for Confucius. Thus, rather than use many specific quotes to exemplify central ideas, I have cited references and summarized key themes and only used occasional quotes to serve as examples. North American career theory and research often is divided into research conducted at two levels of analysis, the organization and the individual, and occasional articles examine the interaction between these two levels. The organizational level of analysis has conceptualized careers as vacancy chains or work-role transitions (White, 1970; Nicholson, 1984), open systems typologies (Sonnenfeld and Peiperl, 1988), or jungle gyms (Gunz, Jalland and Evans, 1998). Each of these perspectives describes how cohorts or groups of people change jobs or roles in an organization as time passes. Mechanisms for sorting people into specific positions include sorting by differences in skills and abilities as well as by relationship networks and organizational policies. Organizational structure, strategy and technology determine the characteristic nature, location, and number of positions.

> it is possible to find significant differences in the pattern of work role transitions made by managers with different mixtures of career jungle gyms....There is a close relationship between a firm's structure, its strategy, and the overall shape of its managers' careers. (Gunz *et al.*, 1998, pp. 25, 34)

The common image is of organizations as vessels full of slots to be filled by successive waves of people. In more recent boundary-less images of organizations, the vessel may have multiple entry points or networks of related chambers and passages between organizations, but career management seen at the level of organizations or inter-organizational networks still focuses on cohort passages in and out of slots located in specific places.

> Thus careers move across firms, rather than within a firm. The film industry provides an empirical base and model for understanding how changes in employment affect careers and inter-firm organization.... The emphasis is on the boundary-less career – defined by movement across the boundaries of separate firms. (Jones, 1996, pp. 58–59)

Scholars working at the individual level of analysis first conceptualized careers in terms of work histories, work interests, and work choices of Euro-American men and women within a single organization (Strong, 1943; Super, 1957; Hughes, 1958; Arthur, Hall and Lawrence, 1989). More recently, scholars have used boundary-less careers, career cycles of motivation, and career networks as images to describe career processes of diverse groups of men and women as they move across organizations and shift emphasis from work to family responsibilities (Arthur and Rousseau, 1996; Agbor-Baiyee, 1997; Iberra, 1997).

Boundary-less career theory is a typical example of career perspectives that address organizational and individual issues at the turn of the millenium. Three chapters of the classic book on boundary-less careers (Arthur and Rousseau, 1996) can serve to illustrate this point of view. In one chapter, Baker and Aldrich (1996) describe boundary-less careers as varying along three dimensions. The first dimension is the number of employers included in a person's work history. The second dimension is the extent to which an individual accumulates knowledge useful in a changing range of career situations. The final dimension is the extent to which employees play an instrumental role in structuring their work histories through their personal identities (authenticity and self-efficacy) and the extent to which their work histories structure their identities.

The career histories of Americans manifest a great deal of heterogeneity across three dimensions: number of employers, extent of knowledge accumulation, and the role of personal identity.... Guided by the life course perspective, we view career process as people's agency – which refers to their ability to make choices and influence their own lives – within the changing structures of constraints and opportunities. (Baker and Aldrich, 1996, pp. 133–4)

In another chapter, Mervis and Hall (1996) extend this perspective on boundary-less careers to consider the phenomenon of psychological success in a protean career. A protean career is a process where someone manages to achieve a variety of experiences, in education, training, work in several organizations, leisure and family life. Individuals face a transition from the old situation where the organization managed its employees' careers to one where people must become self-directed in a risk-filled, changing environment. In this new situation, creating psychological success may require developing commitment to an occupation rather than an organization, building support networks outside of the work setting and otherwise becoming more self-sufficient.

People's core identities may be enlarged by incorporating what we call a commitment to their life's work. In this framework, a person's identity deepens not only through cumulative work experiences but also through work as a spouse, parent, and community member, and especially through work on one's identity. (Mervis and Hall, 1996, p. 252)

A boundary-less career in a project network is how Jones, in a further chapter, describes the work histories and experiences of people in the film industry. This contains several characteristics that many think will become common in boundary-less careers of the twenty-first century (Jones, 1996). Careers are defined as individuals developing a reputation and an array of skills and contacts that enable them to work in a series of projects as sub-contractors. The 'organization' is a complex temporary network of subcontractors offering complex, non-routine, interdependent tasks requiring high levels of novel information processing to complete a single project.

This movement across firms requires individuals to develop skills such as interpersonal communications and knowledge of the industry culture, which facilitate the informal co-ordination of work. Since individuals must constantly compete for projects, one's reputation and informal network contacts are critical for continued employment. Thus individuals have a much greater responsibility for seeking experiences that maintain and extend their skills. (Jones, 1996, p. 72)

Even though this type of career system appears to differ on many dimensions from the earlier twentieth-century bureaucratic career – moving up the managerial ladder in a single organization – it shares certain assumptions with earlier views. (1) Careers extend forward into future time: progression from first to last jobs at the individual level and future labour force utilization at the organizational level. (2) Careers are dynamic – they contain action or motion or change. (3) Careers may be examined at the level of the individual, at the level of one or more organizations, or as a process of the interaction between these levels. (4) The principal causal forces that shape careers are located in individual choice and organizational strategic management processes. (5) Acquiring generalized skills and developing relationship networks are important individual and organizational career mechanisms.

If we compare these assumptions with the assumptions of Confucianism that were outlined earlier, we see that they are similar in most respects but different in several important respects. First, the North American careers literature shares with the Confucian literature the idea that people are becoming as a constant process that proceeds forward in time from birth to death, or at least from first job to last job. This characteristic in the US is often a part of the definition of a career, that is, it is connected to work. The distinction between work and life-as-the-development-of-character is absent from the Confucian ideology. These differences stimulate some new propositions that may generate research useful for firms who want to do business in a Confucian culture. If we use the information generated by this analysis of Confucian thought to suggest alternative points of view for stimulating novel research about careers in the US, the first two propositions that address adult development are:

Proposition 1:  *Effective careers unfold in an ordered process of human development.*
Proposition 2:  *The process of developing a virtuous character is an indicator of career effectiveness.*

Secondly, the assumptions about the appropriate levels of analysis provide a sharp distinction between Confucian and North American thought. Seeing the important categories of human action as occurring in interpersonal dyads, the family and the nation do not easily fit with our current definitions of careers as something that happens at work and work as something that happens in organizations. In North American literature, only occasionally is work career unified with other dimensions of life, such as family member and citizen, in a unified 'Way'.

This difference in assumptions provides one of the key contributions this analysis makes to future career research. In cultures where the work domain is integrated with the other aspects of life, research and theory may need to find new ways of conceptualizing work, family, and citizen careers into a single integrated whole. The theoretical discussions of Mervis and Hall may point the way to a beginning of this journey, but the Western habit of dividing and categorizing in order to label and understand may create a formidable challenge to this effort. Some propositions arising from the Confucian assumptions about these issues include the following:

Proposition 3: *Interpersonal relationships at work are essential units of analysis for careers.*

Proposition 4: *Interpersonal relationships outside of the work domain are essential units of analysis for careers.*

Proposition 5: *Family relationships are essential units of analysis for careers.*

Proposition 6: *The relationships between individuals and the society (or government) are essential units of analysis for careers.*

Proposition 7: *Analysis of careers as a life whole includes the categories of relationships with friends, family, co-workers and government.*

Thirdly, the idea that work is something that happens in a specific institution for an exchange of money, and that work is one of the key things that defines who we are, is inconsistent with Confucian doctrine. In Confucian thought, the life course is the thing that defines us. Employment and income-related concerns are temptations that lure, or that happen as a matter of course, not specific goals to strive for. There is a challenge to Western career assumptions in the Confucian assumption that specific positions will happen as a matter of life course not as an object of individual achievement, self-control, or strategic career planning. In addition, the primacy of work as the central aspect of the meaning of career may well face considerable challenge. The core aspect or meaningful unit of one's life course or career is virtue, according to Confucian thought. This virtue is in the full realization of the individual in his or her social context and network of obligations.

The importance of obligation and propriety in enacting relationships as contrasted with the use of relationships and reputation to get ahead in one's career, is another key difference in assumptions. For career scholars to take the ideas of the Confucian perspective seriously, the dimensions of the network of relationships that one would want to consider vary considerably from traditional North American ideas of networks. A relationship is not valuable to a career for the job opportunities that it brings but for the opportunities offered for a person to demonstrate his or her virtue and true personhood by acting properly, by fulfilling obligations, and by living up to his or her word. These opportunities, if met with care, create a reputation. The reputation is not based on skills but on the virtue of the individuals in one's network of relationships and how virtuously one carries out the obligations of the relationships. One does not enter into a relationship to gain career advantage, but one accepts a relationship with a virtuous person because of what might be learned and strives to live up to that gift and the resulting obligations of the relationship. An individual gains obligations and reputation from relationships, not opportunities.

This perspective suggests that scholars evaluating careers might want to include the dimensions of (a) reputation for being connected to many others who are known to be people of a good character, and (b) the reputation of being trustworthy, harmonious, and of good character oneself, rather than the reputation for having specific job-related skills. These suggestions generate the following:

Proposition  8:  *Career effectiveness requires proper or appropriate behaviour.*
Proposition  9:  *Career effectiveness requires harmonious relationships.*
Proposition 10:  *Reputation as indicated by being in virtuous relationships, and striving to become virtuous, are indicators of career effectiveness.*
Proposition 11:  *The nature of effective career relationships is mutual obligation to enhance the virtue of the other person.*

Finally, the role of learning is central to both Western and Confucian conceptualizations of life, but learning serves very different functions. In the Confucian world, learning is *The Way* not *a way to something else.* If continuous learning is not to keep upgrading job skills to cope with a constantly changing organizational environment, but to continue developing into a truly *human* being, the causal connection of learning and obtaining a job or other rewards is weakened. Or perhaps, the process of becoming a worthy person mediates the causal relationship between learning and employment. Thus:

Proposition 12:  *Career positions occur as a byproduct of seeking to follow a virtuous life rather than the result of developing individual career strategies.*
Proposition 13:  *Career effectiveness requires continuous generalized learning.*

## CONCLUSION

In summary, a comparison of Confucian principles, assumptions, and practices in Chinese organizations with those of contemporary North America reveals several possible additional ideas that career scholars might want to incorporate in their work in the twenty-first century. Scholars might look more closely at the family and the national levels of analysis for conceptualizing careers, or, even more daringly, try to integrate individual, dyadic, family, and citizen aspects of careers into a life career instead of a work career. Rather than developing relationships, reputation, and skills as mechanisms of career success, scholars might explore the role of developing integrity, character and self-awareness in oneself and one's circle of friends as indicators of career success and see what, if any, relationship these have to other aspects of careers.

Perhaps some might say we are talking about apples and oranges even to compare Confucian ideas with Western career ideas because they do not seem to relate to the same things. Confucian literature is about achieving the truly virtuous or human self in relation to others, and the North American career literature is about individuals taking control of or creating their careers in work settings. But it is precisely because they are not about the same things that this interchange can stimulate us to think in new ways about careers for the twenty-first century. If we are curious enough to try to look at the world through different eyes, perhaps we might be able to understand more about our relationships with Asian counterparts and even come to understand more about ourselves. If Western managers want to do business with firms founded in a Confucian society we might be well advised to explore these career propositions based on this philosophy.

## Note

Several translations and commentaries were used in this analysis. Quotes and the identifying book and chapter references (i.e. 1:2) are taken from *The Analects of Confucius: A Philosophical Translation* (Ames and Rosemont, 1998), which incorporates recent archeological findings from the Dingzhou fragments into the text and brings the translation into Modern English. Their attempt to include nuances of the action component of meaning in Chinese characters sometimes produces awkward phrases or unique interpretations. When this occurs, terms used in other translations (usually from Legge, 1893) have been added in parentheses.

*Confucius. Confucian Analects, The Great Learning, and the Doctrine of the Mean: Translated with critical and exegetical notes, prolegomena, copious indexes, and dictionary of all characters,* written in 1893 by James Legge, is one of the first scholarly translations of Confucian sayings into English, and still serves as a benchmark for later translators.

*The Original Analects: Sayings of Confucius and His Successors* (Brooks and Brooks, 1998) uses modern techniques of linguistic scholarship to suggest an order in which the sayings were compiled into *The Analects*. When multiple sayings on a single topic are quoted in this chapter, they have been ordered according to the chronology recommended by Brooks and Brooks.

*Understanding Confucius* (Ding, 1997) was the primary English source on Confucius available in Beijing during 1997–8 and was consulted to identify a contemporary PR Chinese interpretation of Confucius. Comparison of the topics selected for emphasis in this text with the complete works used in other texts illustrate ideas condoned by the PR Chinese government press as worthy to be read by foreigners.

Every effort has been made to contact all the required copyright-holders, but if any have been inadvertently omitted the publishers will be pleased to make the necessary arrangement at the earliest opportunity.

## References

Abegglen, J. C. (1994) *Sea Change: Pacific Asia as the New World Industrial Center* (New York: Free Press).

Agbor-Baiyee, W. (1997) 'A Cyclic Model of Career Motivation', *College Student Journal*, 31(4), pp. 467–72.

Ames, R. T. and Rosemont, H. (1998) *The Analects of Confucius: A Philosophical Translation* (New York: Ballantine Books).

Aoto, Y., Fukuzawa, S., Hoyoshi, R. and Yage, H. (1988) *Nippon, the Land and Its People*, translated by Richard Foster (Tokyo: Nippon Steel Co. Ltd).

Arthur, M. A., Hall, D. T. and Lawrence, B. S. (1989) *Handbook of Career Theory* (New York: Cambridge University Press).

Arthur, M. B. and Rousseau, D. M. (eds) (1996) *The Boundaryless Career: A New Employment Principle for a New Organizational Era* (New York: Oxford).

Baker, T. and Aldrich, H. E. (1996) 'Prometheus Stretches: Building Identity and Cumulative Knowledge in Multi-Employer Careers', in M. B. Arthur and D. M. Rousseau (eds), *The Boundaryless Career: A New Employment Principle for a New Organizational Era* (New York: Oxford) pp. 132–49.

Bell, M., Khor, H. E., Kochlar, K., Ma J. N'guiamba, S. and Lall, R. (1993) *China at the Threshold of a Market Economy* (Washington, DC: International Monetary Fund).

Biggart, N. W. and Hamilton, G. G. (1997) 'Explaining Asian Business Success: Theory No. 4', in M. Orru, N. W. Biggart and G. G. Hamilton (eds), *The Economic Organization of East Asian Capitalism* (Thousand Oaks, CA.: Sage Publications) pp. 97–110.

Brooks, E. B. and Brooks, A. T. (1998) *The Original Analects: Sayings of Confucius and His Successors* (New York: Columbia University Press).

Ding, W. D. (1997) *Understanding Confucius* (Beijing: Chinese Literature Press).

Granrose, C. S. and Chua, B. L. (1996) 'Global Boundary-less Careers: Lessons from Chinese Family Businesses', in M. B. Arthur and D. M. Rousseau (eds), *The Boundaryless Career: A New Employment Principle for a New Organizational Era* (New York: Oxford University Press) pp. 201–17.

Gunz, H. P., Jalland, R. M. and Evans, M. G. (1998) 'New Strategy, Wrong Managers? What You Need to Know About Career Streams', *Academy of Management Executive*, 12(2), pp. 21–37.

Guo, R. (1999) *How the Chinese Economy Works* (London: Macmillan).

Hughes, E. C. (1958) *Men and their Work* (Glencoe, IL: Free Press).

Iberra, H. (1997) 'Paving an Alternative Route: Gender Differences in Managerial Networks', *Social Psychology Quarterly*, 60(1), pp. 91–102.

Jian, B. Z., Shao, X. S. and Hu, H. (1986) *A Concise History of China* (Beijing: Foreign Languages Press).

Jones, C. (1996) 'Careers in Project Networks', in M. B. Arthur and D. M. Rousseau (eds), *The Boundaryless Career: A New Employment Principle for a New Organizational Era* (New York: Oxford University Press) pp. 58–75.

Lee, Ki-baik (1984) *A New History of Korea*, translated by Edward W. Wagner with Edward J. Shultz (Cambridge, MA: Harvard University Press).

Legge, J. (1893) *Confucius. Confucian Analects, The Great Learning, and the Doctrine of the Mean: Translated with critical and exegetical notes, prolegomena, copious indexes, and dictionary of all characters* (Oxford, England: Clarendon Press) (reprinted in 1971 by Dover Books, N.Y.).

Mervis, P. H. and Hall, D. T. (1996) 'Psychological Success and the Boundary-less Career', in M. B. Arthur and D. M. Rousseau (eds), *The Boundaryless Career: A New Employment Principle for a New Organizational Era* (New York: Oxford University Press) pp. 237–56.

Nicholson, N. (1984) 'A Theory of Work Role Transitions', *Administrative Science Quarterly*, 29, pp. 172–91.

Ning, L. (1998) 'Chinese to overtake US GDP by 2015: OECD', *BTOnline*, *Business Times Asia*, 19 October. l.com.sg.

Roberts, K., Kossek, E. E. and Ozeki, C. (1998) 'Managing the Global Workforce: Challenges and Strategies', *Academy of Management Executive*, 12(4), pp. 93–106.

Shen, F. W. (1996) *Cultural Flow between China and Outside World Throughout History* (Beijing: Foreign Languages Press).

Sonnenfeld, J. A. and Peiperl, M. A. (1988) 'Staffing Policy as a Strategic Response: A Typology of Career Systems', *Academy of Management Review*, 13, pp. 588–600.

Strong, E. K. Jr. (1943) *Vocational Interests of Men and Women* (Stanford, CA: Stanford University Press).

Super, D. E. (1957) *The Psychology of Careers* (New York: Harper & Row).

Weidenbaum, M. and Hughes, S. (1996) *The Bamboo Network: How Expatriate Chinese Entrepreneurs are Creating a New Economic Superpower in Asia* (New York: The Free Press).

White, H. C. (1970) *Chains of Opportunity* (Cambridge, MA: Harvard University Press).

Woronoff, J. (1986) *Asian 'Miracle' Economies* (Seoul: Si-sa-yong-o-sa Inc).

Zahra, S. A. (1999) 'The Changing Rules of Global Competitiveness in the 21st Century', *Academy of Management Executive*, 13(1), pp. 36–42.

# 7 Awakening Creative Intelligence for Peak Performance: Reviving an Asian Tradition

Dennis P. Heaton and Harald S. Harung

Frequently, Eastern management looks to the West for new ideas and ways to improve management and performance. Western knowledge has made many valuable contributions to enhance productivity, quality, and performance in general. At the same time it is readily evident – both in the West and the East – that the Western approach has many shortcomings, including growing stress and health problems, widespread lack of happiness, antisocial behaviour, and environmental degradation.

In this chapter we discuss how the East also has something highly valuable to teach the West – the attainment of an inner state of balance and strength as the basis of outer success in harmony with the social and natural environment. We suggest that the widespread practice of selected, scientifically validated meditation techniques, based on ancient Asian wisdom, can enliven the qualities of vitality, health, and holistic wisdom which are needed for the people in any society to achieve progress and prosperity without problems.

The accomplishments and the shortcomings of Western knowledge are attributable to the objective approach of modern science which has provided the ability to manipulate specific laws of nature in an isolated fashion. Incomplete knowledge based on isolated areas of experimental investigation has created unanticipated, harmful side-effects – confirming the saying 'A little knowledge is a dangerous thing'. To complement this fragmented approach of modern science, traditional Indian and Chinese approaches have called for holistic subjective development which establishes the intelligence of the individual in harmony with the organizing intelligence of Nature.

# CONCEPTS OF HUMAN INTELLIGENCE IN ASIAN TRADITIONS

In the context of this volume on human intelligence deployment in Asia, it is appropriate that we define intelligence from the perspective of Asian tradition. So Kam-Tim (1995) provides the following definition of an *intelligent* man, according to the traditional Chinese concept of intelligence:

> One who is capable of displaying more of the creative and intelligence potential of nature or simply living more in tune with the spontaneous tendencies of nature, the nature of Tao, is considered intelligent or wise. A wise man, who is called a 'Sage' or 'Great Man', is one who lives in tune with Natural Law; he knows and lives the creative intelligence of the universe in individual life. (1995, p. 95)

Such a Sage naturally applies his creative intelligence to *accomplish* things in daily life. To illustrate the Chinese understanding that a Sage is man of skilful action, So Kam-Tim quotes His Tz'u: 'In preparing things for practical use and in inventing instruments for the benefit of the world, there are none greater than the Sages' (Legge, 1969, sec. I, ch. 7). He also recounts that '"Inner enlightenment for outer fulfillment" has been the highest ideal of a royal personality of an emperor of China' (p. 96).

Likewise in the Vedic tradition of India, intelligence is understood to be an inherent property of the universe, with which one can align oneself for maximum holistic success. In the introduction to his Vedic University, Maharishi Mahesh Yogi explains that:

> The technique of attuning one's attention to the Constitution of the Universe was given to the ruling dynasties so that the ruler rules from that level of pure intelligence which rules the university: Bhahma Bhavati sarathih – that basic level of creation from where everything emerges and evolves...the total light of Natural Law...the Light of the Self. (1994, p. 225)

Through the application of various technologies from the Vedic tradition, including meditation, Maharishi (1995) explains, it is possible to attain 'administration as problem-free, ever-progressive, and ever-evolutionary as the administration of the universe through Natural Law' (Maharishi Mahesh Yogi, 1995, p. 8).

The strategy of inner enlightenment for outer success is advocated in the traditional Vedic literature, much as it is in the traditions of China. In the following passage it is spoken of as becoming established in Yoga, which can be understood as the integration of individual intelligence and the managing intelligence of the universe:

> Established in Yoga, O winner of wealth, perform actions having abandoned attachment and having become balanced in success and failure, for balance of mind is called Yoga. (Maharishi Mahesh Yogi, 1969, p. 135)

So Kam-Tim (1995) observes: 'Both the Chinese tradition and Maharishi's Vedic Psychology state that intelligence originates from a unified source which is transcendental to space, time, causation, and mental processes. This undivided Oneness of intelligence is called, Tao, Tai Chi, I, or Heaven in the Chinese tradition, or pure intelligence, pure consciousness, cosmic psyche, Self, or Being in Maharishi's Vedic Psychology' (p. 145).

## ENLIVENING HOLISTIC INTELLIGENCE THROUGH MEDITATION

According to Asian tradition, the individual mind can imbibe the orderly and evolutionary intelligence of Nature by gaining a state of stillness. For example, when the Yellow Emperor asks the Sage Juang Ch'eng-tzu for the Tao of government to bring about world peace, the Sage replies:

> The state of Transcendental Being is unmanifest and silent; there is no seeing and no hearing. When we enfold the mind is stillness, the body will naturally become correct. (Chang-tzu, translated by Watson, 1968, ch. 11, p. 4)

Likewise the extraordinary vision of problem-free administration in social systems in Lao Tzu's *Tao teh Ching* is attributed to the practice of 'quietism' to realize the *Tao*, the inherent whole of consciousness (Waley, 1968).

Today, based on extensive scientific research on the physiological, psychological, and sociological benefits of specific meditation techniques, individuals, companies, and governmental organizations in all parts of the world are reviving the Asian tradition of daily meditation practice for growth toward peak performance. In our work as management consultants and researchers, we have been involved with companies utilizing the Transcendental Meditation® (TM®) programme[1] to awaken alertness, vitality, and creative intelligence in their personnel (Harung, 1999).

The Transcendental Meditation (TM) programme was introduced to the world by Maharishi Mahesh Yogi in 1957, and has since then been learned by more than five million people world-wide (Roth, 1987). This technique is described as a natural, effortless, and enjoyable procedure that is practised for 20 minutes twice a day, sitting comfortably in a chair with eyes closed. This practice allows active awareness to gradually settle down so that the silent, unbounded, and unified state of transcendental consciousness can be directly experienced. Transcendental consciousness is a state of *restful alertness* that can be distinguished from the aroused alertness characteristic of waking and from the more inert states of dreaming and sleeping. The settling down of the mind during the transcending process allows the body to gain profound relaxation, which dissolves the stresses blocking the expression of inherent potential. The deep rest and expanded alertness experienced during this procedure establish a basis for dynamic and effective action. A tangible benefit of this stress reduction is that the practice of the TM programme is associated with greatly reduced medical care utilization (Orme-Johnson, 1987; Orme-Johnson and Herron, 1997).

The TM technique is the most thoroughly researched meditation practice in the world today (Murphy and Donovan, 1996). Physiological studies have observed indicators of deep relaxation during the practice of this technique – decreases in respiration rate, heart rate, skin conductance level (a measure of nervousness), and blood lactate (a stress-related hormone) which are greater than those found during simply resting with eyes closed (Wallace, 1993). At the same time, there are hormonal and brain *wave* changes indicating increased alertness during the practice.

Researchers at Stanford University conducted a statistical meta-analysis of 146 independent investigations (Eppley, Abrams and Shear, 1989) and found that the TM programme was about twice as effective as other mental and relaxation techniques in reducing trait (long-term) anxiety. Another meta-analysis found that the TM programme produced twice the effect size of other meditation and relaxation techniques in growth of self-actualization (Alexander, Rainforth and Gelderloos, 1991).

Based on holistic conceptions of intelligence in Chinese and Vedic traditions, So Kam-Tim (1995) investigated the development of holistic intelligence through the Transcendental Meditation programme in 3 randomized, blind, controlled studies (6–12 months) conducted with 353 Chinese students. He found that practice of the TM technique developed significantly greater creativity, practical intelligence, reduced anxiety, neural efficiency, and field independence – all of which are indicators of intelligence in the body, mind, and behaviour. So Kam-Tim's findings are consistent with prior related research findings that the TM technique promotes the growth of fluid intelligence and practical intelligence (Cranson *et al.*, 1991), field independence (Dillbeck, Assimakis, Raimondi, Orme-Johnson and Rowe, 1986) and creative thinking (Travis, 1979).

Another major study of holistic development through the TM technique was conducted by Chandler (1991) and Chandler, Alexander and Heaton (in press). This study used three convergent measures to look at the long-term effects of meditation practice in a 10-year longitudinal study. The findings were that meditators grew in intimacy motivation (a measure of positive, caring emotions), principled moral reasoning, and ego or self-development. The measure of self-development is itself considered an indicator of holistic psychological maturity, encompassing self-concept, intellectual analysis and synthesis, and interpersonal relations (Loevinger, 1976; Alexander *et al.*, 1990; Kegan, 1994). A milestone in this growth is when the individual becomes post-conventional – able to think and act independent of the collective mentality of his surrounding social system. It is striking that at post-test 53 per cent of the meditating subjects achieved post-conventional levels of development, compared with less than 10 per cent in the control group and in most other samples of adult subjects. In fact, the meditating group achieved the highest proportion of post-conventional ego development yet recorded among adolescent and adult samples – including Harvard University alumni and senior management samples. The significance of this finding to performance in business settings is discussed in the section below.


## SIGNIFICANCE OF PSYCHOLOGICAL DEVELOPMENT FOR PERFORMANCE

The types of psychological changes brought about by the TM technique have direct and practical consequences for performance in business. In this section we elaborate on the significance of one particular finding – that the TM programme has been seen to promote post-conventional ego develop-ment. In various models of the higher ranges of psychological development (e.g., Loevinger, 1976; Torbert, 1991; Kegan, 1994), post-conventional stages are associated with greater capacity for breakthrough thinking, mutual respect, and ethical values. The promotion of higher levels of psychological development, then, is understood to be a fundamental approach to manage-ment development (Merron, Fisher and Torbert, 1987) that can be applied in any industry. Management researchers have identified that personnel in organizations are operating at a wide range of levels of psychological devel-opment. Table 7.1 summarizes six studies which identified the developmental stages of a total of 500 managers.

*Table* 7.1   Stages of ego development

| Stage name | % at stage | Description |
| --- | --- | --- |
| Opportunist | 2 | Short-time horizon, concrete, distrustful, manipulative |
| Diplomat | 8 | Conforms with rules and group norms, thinks in stereotypes, feels shame if violates norm |
| Technician | 45 | Interested in problem-solving, efficiency over effectiveness, perfectionist, evaluates self and others based on craft logic |
| Achiever | 36 | Results-oriented, long-term goals, initiative, respects differences, seeks mutuality rather than hierarchy in relationships, open to feedback |
| Strategist | 9 | Ability to reframe situations and define new goals, role flexibility, creative conflict resolution, concern with total organization in the environment, empowers others |

The first developmental stage in Table 7.1 is *opportunist*. Opportunists have a short-term horizon and the inability to receive feedback and assume responsibility (externalizing blame). Next, the *diplomat* identifies with others' expectations. This stage is typically found in the earlier teen years, though 8 per cent of the adult managers were at this stage. A *technician's* frame has shifted from the expectations of others, which are found to be multiple and conflicting, to dedication to the 'craft logic' of a single field of endeavour, e.g., engineering, accounting, marketing, law. It is a challenge for the technician to understand people with other perspectives and objectives. As Table 7.1 shows, technician is the most common developmental position – representing 45 per cent of adults sampled. *Achiever* is the second most common developmental stage among managers tested. Compared to the technician stage, the achiever is results-oriented, inspiring, respectful of individual differences, open to feedback, and seeks mutuality rather than hierarchy in relationships. Yet, achievers are still limited to operating *within* a conventional frame and restricted to the implementation of an *existing* strategy, rather than the identification of more value-adding goals and the creation of new, more productive systems.

At the *strategist* stage, a person ceases to take the frame of the existing social system for granted and becomes interested in what a best system would be. The improvement of performance at this post-conventional stage involves a recognition that there may not only be a need to change actions, but that a change of goals, structures, and human values may also be needed. Strategists are better able to integrate intellect and emotions, to handle stress, to act from a distinctive set of visions and values, and to be genuinely appreciative of differences of perspective.

Still beyond the strategist stage is a stage of personality development which Loevinger called *integrated* and Maslow called *self-actualizatized* (Maslow, 1968). The integrated phase is characterized by the resolution of inner conflicts (Loevinger, 1976), increased subjective freedom; broader, more comprehensive and integrative awareness; and sound mental health. According to Maslow (1968), self-actualized human beings have:

- Superior perception of reality
- Increased acceptance of self, of others, and of nature
- More confidence and positive philosophy about life
- Spontaneity in thought and behaviour
- Greatly increased creativity
- Strong empathy for others
- Concern with higher-order human values such as truth, beauty, and justice
- Ability to look at life from an objective viewpoint

From an analysis of over 3,000 subjects, it has been concluded that less than 1 per cent were at an integrated or self-actualized stage (Cook-Greuter, 1994).

## Greater Capacity for Business Expansion at Higher Development Stages

Studies indicate that higher stages of development are associated with differences in business performance. For example, a study of entrepreneurial physicians managing their own practices found:

> The technicians – insisting on hands-on participation in every technical phase of their operations – are able to see essentially one patient at a time. The achievers – delegating significant aspects of the operation to their staffs, with oversight – can see essentially three patients at a time. The strategist – able to see critical gaps in services, move into unoccupied niches, and create contracts that motivate partner physicians – are able to create multi-site practices and see three times again as many patients. (Torbert, 1991)

*Table* 7.2  Benefits of the TM technique in business settings

- Enhanced employee effectiveness
- Improved job performance
- Improved organizational contribution
- Enhanced job satisfaction
- Improved holistic thinking in managers
- Enhanced receptivity to important company issues
- Improved alertness and ability to achieve more with less
- Improved tolerance and self-confidence
- Increased psychological well-being
- Improved general health
- Decreased cigarette and hard liquor use
- Greater ability to manage work-related stress
- Reduced physiological arousal
- Reduced trait anxiety, job tension, insomnia, and fatigue
- Improved relations with others

## Developing Toward Greater Capacity for Intercultural Competence

Individuals at the post-conventional strategist and integrated stages may be better candidates for intercultural assignments because they are better able to appreciate and orchestrate the diverse perspectives of multiple parties. The term 'post-conventional' refers to their ability to orient themselves from an inner source of creative intelligence, rather than depend upon familiar patterns of thought and behaviour. This gives them greater capacity to adapt and learn in the complex, fast-changing, and ambiguous situations that characterize today's international business scene.

## The TM Programme in Business Settings Throughout the World

Above we reviewed some of the hundreds of research studies on the psychological and physiological effects of the TM technique, with particular emphasis on its role in stimulating psychological developmental in adulthood. Now let us look specifically at the results in companies in Asia, North America, and Europe which have implemented the Transcendental Meditation technique for human resource development (see review by Schmidt-Wilk, Alexander and Swanson, 1996). Table 7.2 summarizes selected research findings in business settings.

In Japan, a study involving 800 employees in a major manufacturing corporation found considerable health benefits among TM practitioners compared with controls: less health complaints, smoking, insomnia, and psychological distress (Haratani and Henmi, 1990). A Japanese executive from Toyota noted: 'Now I always feel relaxed and I can maintain equilibrium of mind and stability when in a strained work situation' (quoted in Harung, 1999, p. 70). The utilization of the TM technique in Indian corporations is mentioned in press articles published in India, such as in *Businessworld* (1995) and *Economic Times of India* (Kumar, 1997).

In the USA, Alexander *et al.* (1993) measured the impact of the Transcendental Meditation technique in two settings in the automotive industry: a cluster of manufacturing plants owned by a Fortune 100 company, and a small sales distribution company. Those who regularly practised the TM technique demonstrated enhanced employee effectiveness, job satisfaction, and work/personal relationships. In another study, managers practising this technique at the Midwestern US headquarters and nearby facilities of a well-established medical equipment developer and manufacturer improved significantly on 15 measures of vitality, physical complaints, and healthful behaviours (DeArmond, 1996). Moreover, observers rated the TM practitioners higher than non-practitioners on psychological well-being and organizational contribution.

A recent study of workers and managers in a small US company found that individuals who practised the Transcendental Meditation technique grew significantly in their expression of leadership behaviours when compared with other workers from the same company. The TM practitioners' growth spanned all five categories of the Leadership Practices Inventory (Kouzes, Posner and Peters, 1996): encouraging the heart, enabling others to act, modelling the way, challenging the process, and inspiring a shared vision. Also, when interviewed, the meditators described increased comfort in taking initiative, increased ability to negotiate, increased ability to think clearly, increased energy, and decreased tendency to be affected by stress (McCollum, 1999).

A case-study of a Norwegian top management team found that the directors had become 'more open-minded', 'more patient', 'more positive', 'more awake in the meetings', and 'more in control of their emotions, more stable, not quiet, but less ups and downs' (Schmidt-Wilk, Alexander and Swanson, 1995). An investigation in Sweden found that a group of meditating managers in a private business grew in holistic thinking and became more aware of the important issues facing the company, as assessed by a superior (Gustavsson, 1992, p. 31). Moreover, the practice of the TM technique develops one's alertness and holistic creative intelligence for managerial decision-making, as has been aptly expressed by this American CEO:

After meditating, I have the mental clarity and alertness for laser-like focus on the details, and, at the same time, for broad comprehension so I don't get lost in the details. I find myself continuously growing in insight and intuition, as well as in the ability to focus and analyze. Over my years in business, the TM technique has been a real competitive advantage. (quoted in Harung, 1999, p. 70)

Companies which have offered the TM technique for their human resources have found that the culture and performance of the company as a whole benefit. A medium-sized chemical manufacturing company supplying the US automotive industry had survived four years of flat sales and declining profits in a tightening market. A new CEO offered the Transcendental Meditation programme to the workforce as a turnaround strategy. Fifty-six of the 70 employees learned the technique. As a result of this voluntary programme, records showed increases in productivity and profitability. Net income increased steadily as the percentage of participating employees rose over a six-year period. Productivity as measured by units produced per man-hour increased by 52 per cent, and annual sales per employee increased by 88 per cent. At the same time, labour costs as a percentage of sales revenues decreased 39 per cent, the number of work days lost due to poor health or injuries declined more than 50 per cent, and absenteeism declined 89 per cent. These improvements were greater than corresponding improvements within the auto industry. Management and sales staff became more accessible to one another, employees felt freer to share their ideas and insights, and communication patterns became more open (Schmidt-Wilk *et al.*, 1996).

In Germany, the directors of a finance company offered the Transcendental Meditation programme to employees as part of their in-service training. After 20 of the 100 employees had been instructed in the programme, the directors noted a general improvement in the work climate. The directors also reported improvements in key performance areas: reduced absenteeism due to illness, a marked reduction in customer complaints, an increased volume of lending, and a significant decrease in the number of insolvency cases. Four years of radical growth followed, in which balance-sheet totals grew by 230 per cent and profits increased over 300 per cent with only a 28 per cent growth in personnel (Gottwald and Howald, 1989, reviewed in Schmidt-Wilk *et al.*, 1996).

## CREATIVE INTELLIGENCE AND PEAK PERFORMANCE

Beyond the psychological development described by contemporary Western social science, Vedic Psychology describes a still more advanced range of human development with the potential to transform individual and organizational action toward permanent peak performance (Harung, 1999). The foundation of this advanced range of development is the growing experience of transcendental consciousness, a state of silent wakefulness in which individuals are experiencing the deepest aspect of themselves. Maharishi Mahesh Yogi describes transcendental consciousness as:

> A state of inner wakefulness with no object of thought or perception, just pure consciousness aware of its own unbounded nature. It is wholeness, aware of itself, devoid of differences, beyond the division of subject and object – transcendental consciousness. It is a field of all possibilities, where all creative potentialities exist together…but as yet unexpressed. It is a state of perfect order, the matrix from which all the laws of nature emerge, the source of creative intelligence. (1976, p. 123)

Through repeated experience of transcendental consciousness, this source of creative intelligence becomes increasingly available in active living, and behaviour becomes more free, more blissful, and more powerful.

Vedic Psychology explains that the individual practising the Transcendental Meditation technique grows toward peak performance by spontaneously expressing more fully the principle of least action which is evident in all the laws of nature. Then action flows without friction, with 'the minimum amount of energy expended and with the maximum amount of work…the least strain and the maximum amount of gain to the doer and to the surroundings' (Maharishi Mahesh Yogi, 1969, p. 151).

This theme of 'least action' in natural peak performance has also been highlighted by leading thinkers in the West. In his Presidential Address to the Seventh World Productivity Congress, Tor Dahl stated:

> Peak performance is often beautifully simple: there is no waste of motion or energy. There is knowledge and focus. There is continuity – hence no waiting or downtime. There is an elegance that transcends and transforms. All of us seek it in all that we do, and when we succeed, there is a satisfaction so deep and abiding that our spirits are lifted and our minds are at peace. (Dahl, 1990)

Abraham Maslow made a similar observation regarding the creativity of self-actualized people – that they make an 'operation more neat, compact, simpler, faster, less expensive, turning out a better product, doing with less parts, a smaller number of operations, less clumsiness, less effort, more foolproof, safer, more "elegant", less laborious' (quoted in Garfield, 1986).

A business example that illustrates the principle of action in accord with the principle of least action, is the investment success of Berkshire Hathaway chairman Warren Buffet – one of the wealthiest people in the world. Buffet has learned that good businesses enable the investor to make an easy decision, but tough businesses require difficult decisions. If the decision to purchase a business is not easy, he will not pursue the company (Hagstrom, 1997).

To examine the relationship between peak performance and states of heightened consciousness predicted by Vedic Psychology, we conducted a preliminary study involving 22 world-class performers – people known internationally for their ability to achieve and maintain a position among the top performers in professions including business, government, education, and the performing arts (Harung, Heaton, Graff and Alexander, 1996). We found that world-class performers reported more frequent experiences of silent wakefulness on its own, inner silence coexisting with outer activity, and more 'fortunate coincidences' (evidence of achievement through least action) than have been found among other, less-distinguished, performers.

Furthermore, an examination of world-leading performers within science (e.g., Einstein, Maxwell, Schrödinger), music (e.g., Mozart, Beethoven, Bach), government (e.g., former Egyptian president Anwar El Sadat), and sports (e.g., Pelè, Billie Jean King, Roger Bannister) shows that they all have glimpses of higher states of consciousness (Pearson, in press). For example, one of the history's greatest tennis players, Billie Jean King, who captured a record twenty Wimbledon titles and elevated women's athletics to new heights, reports the following glimpse of the inner freedom of higher states of consciousness:

> It almost seems as though I'm able to transport myself beyond the turmoil on the court to some place of total peace and calm. Perfect shots extend into perfect matches.... I appreciate what my opponent is doing in a detached abstract way. Like an observer in the next room.... It is a perfect combination of [intense] action taking place in an atmosphere of total tranquillity. When it happens I want to stop the match and grab the microphone and shout that's what it's all about, because it is. It's not the big prize I'm going to win at the end of the match or anything else. (King and Chapin, 1974, p. 199)

Even though she is performing with great dynamism, she feels in 'some place of total peace and calm', 'like an observer in the next room'. Consistent with the predicted association of spontaneous right action with the growth of consciousness, she is performing at her peak: 'perfect shots extend into perfect matches...perfect combination of action taking place in an atmosphere of total tranquillity'.

Peak performing individuals have experienced that their finest actions are computed by the all-pervading cosmic intelligence of nature, not just their individual intellect and feelings. The German composer Johannes Brahms described how creativity spontaneously flows as a result of connecting to cosmic power: 'The powers from which all truly great composers like Mozart, Schubert, Bach, and Beethoven drew their inspiration is the same power.... The real genius draws on the Infinite source of Wisdom and Power... because they [have] linked themselves to the infinite energy of the Cosmos' (Pearson, in press, p. 261). Brahms recounted how this elevated experience naturally leads to peak performance in composing: 'Straight away the ideas flow in upon me ... and not only do I see distinct themes in my mind's eye, but they are clothed in the right forms, harmonies, and orchestrations. Measure by measure, the finished product is revealed to me when I am in those rare, inspired moods' (Pearson, in press, p. 260).

This evidence from peak performing individuals is consistent with the theory of Vedic Psychology that the creative intelligence at the silent, transcendental level of the mind is the foundation of peak performance. The demonstrated effectiveness of regular practice of the Transcendental Meditation technique and the advanced TM-Sidhi® in cultivating creative intelligence holds promise that peak performance, rather than being a rare exception, can be systematically developed on a wide-scale basis.


## CONCLUSION

In today's global business environment, organizations throughout the world are experimenting with new practices of management and organization in the attempt to improve individual and organizational performance. Yet, evidence suggests that 80 per cent of all initiatives to fundamentally improve the performance of organizations fail (Senge, 1994; Jones, 1995; McMaster, 1996). Despite the good intentions of practitioners, consultants, authors, and teachers, the needed performance improvements have not occurred, especially in knowledge and service jobs. In our analysis, the primary constraint to performance improvement has been the fact that the vast majority of adults are functioning at developmental levels well below our full human potential.

While Western social science recognizes the significance of psychological development for performance, it has yet lacked the knowledge to practically effect developmental advances in adults. By contrast, the Transcendental Meditation technique, from the Vedic tradition, provides an effective methodology for promoting developmental transformations. An accumulating body of scientific evidence suggests that the TM technique brings about a variety of desirable psychological and physiological changes in the individual, including personality growth, moral development, improved health, growth of creative thinking, and improved performance on tests of intelligence.

The use of scientifically validated meditation techniques in businesses represents a revival of an Asian tradition of subjective technologies for holistic development. Such systematic attention to subjective development, we feel, is essential now for people in all parts of the world. The spread of Western science, technology, and lifestyles is impinging on the integrity of cultures and ecosystems around the world and needs to be balanced by the Asian contribution of awakening inner creative intelligence. Human resource development programmes for the global citizen of the twenty-first century can incorporate the best of both the West and the East to achieve progress without problems – outer success and inner harmony.

## Note

1.  ®Transcendental Meditation, TM, and TM-Sidhi are registered trademarks licensed to Maharishi Vedic Education Development Corporation and used under sub-licence.

## References

Alexander, C. N., Davies, J. L., Dixon, C., Dillbeck, M. C., Druker, S. M., Oetzel, R., Mühlman, J. M. and Orme-Johnson, D. W. (1990) 'Growth of Higher Stages of Consciousness: Maharishi's Vedic Psychology of Human Development' in C. N. Alexander and E. J. Langer (eds) *Higher Stages of Human Development: Perspectives and Adult Growth* (New York: Oxford University Press) pp. 286–341.
Alexander, C. N., Rainforth, M. V. and Gelderloos, P. (1991) 'Transcendental Meditation, Self-Actualization, and Psychological Health: A Conceptual Overview and Statistical Meta-Analysis', *Journal of Social Behavior and Personality*, 6, pp. 189–247.
Alexander, C. N., Swanson, G. C., Rainforth, M. V., Carlisle, T. W., Todd, C. C. and Oates, R. (1993) 'Effects of the Transcendental Meditation Program on Stress-Reduction, Health, and Employee Development: A Prospective Study in Two Occupational Settings', *Anxiety, Stress, and Coping*, 6, pp. 245–62.
*Businessworld* (1995) 'Meditation is the New Mantra', *Businessworld*, 4–17 October.
Chandler, H. M. (1991) 'Transcendental Meditation and Awakening Wisdom: A Ten-Year Longitudinal Study of Self Development', *Dissertation Abstracts International*, 51(10B), p. 5048.
Chandler, H., Alexander, C. N. and Heaton, D. P. (in press) 'Transcendental Meditation and Postconventional Self Development', *Journal of Social Behavior and Personality*.
Cook-Greuter, S. (1994) 'Rare Forms of Self Understanding in Mature Adults', in M. E. Miller and S. R. Cook-Greuter (eds), *Transcendence and Mature Thought in Adulthood: The Further Reaches of Adult Development* (Lanham, MD: Rowman & Littlefield) pp. 39–70.

Cranson, R. W., Orme-Johnson, D. W., Gackenbach, J., Dillbeck, M. C., Jones, C. H. and Alexander, C. N. (1991) 'Transcendental Meditation and Improved Performance on Intelligence-Related Measures: A Longitudinal Study', *Journal of Personality and Individual Differences*, 12, pp. 1105–16.

Dahl, T. (1990) 'Creating Lasting Change: A Presidential Address', Presented at the Seventh World Productivity Congress, Kuala Lampur, Malaysia, 19 November.

DeArmond, D. L. (1996) 'Effects of the Transcendental Meditation Program on Psychological, Physiological, Behavioral and Organizational Consequences of Stress in Managers and Executives, Doctoral Dissertation, Maharishi University of Management. *Dissertation Abstracts International*.

Dillbeck, M. C., Assimakis, P. D., Raimondi, D., Orme-Johnson, D. W. and Rowe, R. (1986) 'Longitudinal Effects of the Transcendental Meditation and TM-Sidhi Program on Cognitive Ability and Cognitive Style', *Perceptional and Motor Skills*, 62, pp. 731–8.

Drath, W. H. (1990) 'Managerial Strengths and Weaknesses as Functions of the Development of Personal Meaning', *Journal of Applied Behavioral Sciences*, 26, pp. 83–499.

Eppley, K. R., Abrams, A. I. and Shear, J. (1989) 'Differential Effects of Relaxation Techniques on Trait Anxiety: A Meta-Analysis', *Journal of Clinical Psychology*, 45, pp. 957–74.

Garfield, C. (1986) *Peak Performers: The New Heroes of American Business* (New York: Morrow).

Gottwald, F. T. and Howald, W. (1989) *Selbsthilfe durch Meditation* [Self-help through meditation]. (Munich, Germany: Moderne Verlagegesellschaft).

Gustavsson, B. (1992) 'The Transcendent Organization', Doctoral Dissertation, Stockholm, Sweden: University of Stockholm, Department of Business Administration.

Hagstrom Jr., R. G. (1997) *The Warren Buffet Way* (New York: John Wiley).

Haratani, T. and Henmi, T. (1990) 'Effects of Transcendental Meditation (TM) on Health Behavior of Industrial Workers', *Japanese Journal of Public Health*, 37, 10, p. 729.

Harung, H. S. (1999) *Invincible Leadership: Building Peak Performance Organizations by Harnessing the Unlimited Power of Consciousness* (Fairfield, IA: Maharishi University of Management Press).

Harung, H. S., Heaton, D., Graff, W. and Alexander, C. N. (1996) 'Peak Performance and Higher States of Consciousness', *Journal of Managerial Psychology*, 11, pp. 3–23.

Jones, S. (1995) 'The Democratic Dimension of Quality, Innovation and Long-Term Success', *The TQM Magazine*, 7, pp. 36–41.

Kegan, R. (1994) *In Over Our Heads: The Mental Demands of Modern Life* (Cambridge, MA: Harvard University Press).

King, B. J. and Chapin, K. (1974) *Billie Jean* (New York: Harper & Row).

Kouzes, J. M., Posner, B. Z. and Peters, T. (1996) *The Leadership Challenge: How to Keep Getting Extraordinary Things Done in Organizations*, 2nd edn (San Fransisco: Jossey-Bass).

Kumar, M. (1997) 'Yogi's Corporate Mantra', *The Economic Times of India*, 25 September.

Legge, J. (1969) *I Ching, Book of Changes* (New York: Bantam Books).

Loevinger, J. (1976) *Ego Development: Conceptions and Theories* (San Francisco: Jossey-Bass).

Maharishi Mahesh Yogi (1969) *On the Bhagavad-Gita: A New Translation and Commentary*, Chapters 1–6 (Baltimore, Maryland: Penguin).

Maharishi Mahesh Yogi (1994) *Vedic Knowledge for Everyone: Maharishi Vedic University Introduction* (Vlodrop, Holland: Maharishi Vedic University Press).

Maharishi Mahesh Yogi (1995) *Maharishi University of Management: Wholeness on the Move* (Vlodrop, Holland: Maharishi Vedic University Press).

Maslow, A. H. (1968) *Toward a Psychology of Being*, 2nd edn (New York: Van Nostrand Reinhold).

McCollum, B. (1999) 'Leadership Development and Self Development: A Theoretical and Empirical Exploration', Doctoral Dissertation, Maharishi University of Management, *Dissertation Abstracts International*.

McMaster, M. D. (1996) *The Intelligence Advantage: Organizing for Complexity* (Newton, MA: Butterworth-Heinemann).

Merron, K., Fisher, D. and Torbert, W. R. (1987) 'Meaning Making and Management Action', *Group and Organization Studies*, 12, pp. 274–86.

Murphy, M. and Donovan, S. (1996) *The Physical and Psychological Effects of Meditation: A Review of Contemporary Research with a Comprehensive Bibliography 1931–1996*, 2nd edn (Sausalito, CA: Institute of Noetic Sciences).

Nidich, S. I., Ryncarz, R., Abrams, A., Orme-Johnson, D. W. and Wallace, R. K. (1983) 'Kohlbergian Cosmic Perspective Responses, EEG Coherence, and the Transcendental Meditation and TM-Sidhi Program', *Journal of Moral Education*, 12, pp. 166–73.

Orme-Johnson, D. W. (1987) 'Medical Care Utilization and the Transcendental Meditation Program', *Psychosomatic Medicine*, 49, pp. 493–507.

Orme-Johnson, D. W. and Herron, R. (1997) 'An Innovative Approach to Reducing Medical Care Utilization and Expenditures', *American Journal of Managed Care*, 3, pp. 135–44.

Pearson, C. (in press) The Supreme Awakening: *First-Hand Glimpses from People Throughout History of Awakening to the Inner Reality of Life* (Fairfield, IA: Maharishi University of Management Press).

Roth, R. (1987) *Maharishi Mahesh Yogi's Transcendental Meditation*, rev. edn (New York: Donald I. Fine, Inc.).

Schmidt-Wilk, J., Alexander, C. N. and Swanson, G. C. (1995) 'Introduction of the Transcendental Meditation Program in a Norwegian Top Management Team', in B. Glaser (ed.), *Grounded Theory: 1984–1994* (Mill Valley, CA: Sociology Press) pp. 563–87.

Schmidt-Wilk, J. Alexander, C. N. and Swanson, G. C. (1996) 'Developing consciousness in Organizations: The Transcendental Meditation Program in Business', *Journal of Business and Psychology*, 10, pp. 429–44.

Senge, P. M. (1994) *The Fifth Discipline: The Art and Practice of the Learning Organization*, 1st edn (New York: Doubleday).

So Kam-Tim (1995) 'Testing and Developing Holistic Intelligence in Chinese Culture using Transcendental Meditation', Doctoral Dissertation, Maharishi University of Management. *Dissertation Abstracts International*.

Torbert, W. R. (1991) *The Power Of Balance: Transforming Self, Society, and Scientific Inquiry* (Newbury Park, CA: Sage).

Travis, F. (1979) 'The TM Technique and Creativity: A Longitudinal Study of Cornell University Undergraduates', *Journal of Creative Behavior*, 13, pp. 169–80.

Waley, A. (1968) *The Way and its Power* (London: George Allen & Unwin).

Wallace, R. K. (1993) *The Physiology of Consciousness* (Fairfield, IA: Maharishi International University Press).

Watson, B. (1968) *Chuang-tzu* (New York: Columbia University Press).

# 8 Consciousness as the Basis of Cognition, Culture and Practice in Pre- and Post-Crises Asia

Mohan Raj Gurubatham

The format of this chapter takes us through a management consultant's day-to-day life in Malaysia through a series of cases. These are based upon actual experiences but names of people and institutions have been changed. The events in the story are then analyzed through theoretical and meta-theoretical discussions in the form of 'reflections' to explain the events described in the cases. These explanations, while drawn from theory, are admittedly in-and-of themselves speculative, but nevertheless they draw attention to the yawning chasm in the management literature, as they are dominated by Western thought and constructs. Given the paucity of empirical research in 'post-crises' Asia, it is hoped that the issues raised in this chapter will forge a new agenda for rigorous research by identifying new dimensions of personal experience and cultures with a finesse that needs to precede quantification. In other words, we recognize we need to qualify before we quantify. The conclusion raises these new directions in research and urges the deeper spiritual understanding of culture within the context of consciousness.

We begin this chapter with a scene-setting story – later we will reflect on the underlying processes and behavioural drivers.

I

The new elevated trains that ply through modern Kuala Lumpur (KL) do so every day with clockwork precision. They are driverless but automatic. Commuters stand and sit as outside the scooters, cars, buses, trucks and changing landscapes whizz by. Sunil David is a consultant: he was born and raised in Malaysia, educated in the UK and America, and he now works for a major strategy and consulting firm with offices all over the world. The office that he is based in is in Kuala Lumpur and was opened recently, just before the Asian economic crisis in 1997. He is on his way to an appointment with a client.

The client is a major bank that has merged to become the second largest bank in the country. The person he is to meet is an expatriate director of the new bank. Sitting in the train, Sunil is half-aware to the unfolding spectacle of skyscrapers, high-rises, minarets, temples, churches and domes. These, along with advertising billboards portraying the good life of cosmopolitan chic, punctuate Kuala Lumpur's landscape. They appear and disappear as if someone cannot make up their mind to fast-forward or rewind a video of Asian history. The landscape is bristling with the towers of modern commerce and condominiums. It appears overall as quiet Asian pride resplendent of a new and prosperous Malaysia. There is a sooty drizzle in the air, yet he is cocooned from the noise of the traffic and the lackadaisical street manners.

Sunil is of Indian parentage, Western education, and practises an 'Eastern spirituality'; though he was raised as a Protestant by his parents. He can't help but reflect on his own culture, personality and ethnicity. He takes a glance at his watch and finds it is a quarter-to-nine a.m. He knows that he will have ample time to make his appointment. He notices that the passengers who are seated and standing around him are wearing an assortment of timepieces. There is a medley of Casios, Swatches, the occasional Rolex and other leading Swiss brands. His own watch has an 'old' analogue face. Next to him, an older man wears a digital watch. The digital watch shows the time to be precisely 8:47 a.m. Sunil needs to take a little extra effort to read this exact rendition of time. With this watch the man appears to be task oriented, and somehow youthful.

Sunil alights from the train clutching his satchel and the newspaper which he did not want to read on the train. The sooty drizzle is now found to be unpalatable. He notices that the newspaper has a report on the smog, attributing it to the 'discourtesy' of forest burning in neighbouring Sumatra, in Indonesia. To get to the bank, he waits by the pedestrian traffic lights to cross the street. He looks for an 'API' (Air Pollution Index) quotation in the newspaper, but cannot find one. Most people seem to wait by the kerb for the pedestrian light to change to green, but some brave the red light and attempt to cross the street. The traffic seems to have an informal organization which violates lane discipline at many points. As he arrives for the meeting, Sunil notices the titles of the Board of Directors displayed prominently on the plaque. Some of the titles, apart from doctorates, are state endowed. They represent a title conferred by the state or the king for recognition of outstanding service to the nation.

Sunil is ushered into the boardroom. He waits for the Director to appear. The secretary offers him a drink; he asks for tea. The director then appears, and greets Sunil with a very firm handshake. He asks about Sunil's trip, and then tells him to sit down. The director, Tan Sri Brunner, is a Swiss expatriate who has lived in Malaysia for over 20 years. The salutation 'Tan Sri' is a title conferred by the King of Malaysia recognizing Brunner's long tenure and commercial work in the country. This title is extremely prestigious and has its roots in feudal medieval times. The director has just joined the bank which resulted from a major merger of two very traditional and distinct Asian cultures. A few more senior managers now appear, and they begin to seat themselves respectfully around the large boardroom table. Sunil replies that he took the train. Brunner affirms that it was a smart move since he does not trust the KL traffic which is growing at the rate of 8–10 per cent a year. Brunner is of course chauffeur-driven.

It is now 9:15 a.m., and Brunner asks where the rest of the senior managers are. 'They will be here' is the reply, 'traffic you know...', and 'Malaysian time...'. Malaysian time is an endearing and enduring euphemism for being late. The others arrive and the meeting starts. There is a careful but warm formality in the introductions, and Brunner sits at the head of the table. He complains that the room should be brighter even with the lights on. Someone suggests that the window blinds be drawn open, and they briskly do just that while Brunner sits.


**Reflection**


*Power Distance and Hierarchy*

In the now classic treatise of modern management sociometrics, Geert Hofstede (1980), in his protean exposition of 'power distance' explains that 'low power' cultures such as the United States, Canada, the United Kingdom, Australia and Israel are depicted as 'looking for equality'. There is a markedly pragmatic approach to status where organizational hierarchy is relatively unimportant. These broadly Western nations, to be sure, have variations amongst them but overall they may be characterized as 'low power distance'.

Malaysia in contrast has the highest power distance, and shares this cultural trait with other East Asian nations, who uphold similar high levels of this index (Hofstede, 1991a, 1991b). High power distance is the 'cultural programming' that drives thinking and behaviour. There is respect for hierarchy that is manifest in attending to and addressing titles, salutations, and deferring judgments to, and not openly contradicting, superiors. Deference to authority and acquiescence is very prevalent in a high power distance culture. Hence the formality in introductions, seating arrangements and adherence to salutations noticed in the case above.

## Context and Relationships

One might note that the study of communication throughout Japan, China and also in Mediterranean (Latin) cultures invokes consideration of the great seminal work by Edward Hall (1976) – which is related to the work of Hofstede. Yet less is known about Malaysia and South East Asia in Hall's continuum in hard empirical terms. Hall coined the concept of high context (HC) cultures vs. low context (LC) cultures.

Basically in HC cultures communication processes are non-verbal, non-linear and 'analogue'. Analogue representations are patterned, relational information and involve a compromise between the abstract and the concrete codes involved in mental processing. They invoke psychological sets. Further, analogue processing is quick. Relationships in HC are also 'warm' (Berry *et al.*, 1992, p. 42), and in the field of business, relationships are understood to be important. Time spent on relationship building may be prioritized over tasks, and it is common for punctuality to be traded off towards relationship building. However, there are differences among Asian cultures, with some being perceived as more warm than others. From descriptions, and long exposure to Asian cultures, experience suggests that East Asian and Indian cultures are high context. As such, HC appears to be an artifact of pre-industrial societies. These societies have extended families and they emphasize relationships wherein time is taken to build and nurture relationships. What is 'said' in high context cultures, involves not just speech or writing, but also the nuances of feeling, spatial information and relationships.

Not reporting the air pollution index (API) reading in the mass media is considered good manners in the social context of nation building. Mass media are integrated into the wider social context and government agenda which in this case desires to maintain public morale. The newer context is of a nation recovering from a spate of setbacks such as the economic crisis, and the smog which had brought about billions of dollars of losses to the region in 1998. What is not said is as communicative as what is said, if not more subtle and powerful.

## Cognitive Bases of Context

Analogue representations appear to be more rapidly processed through analogue codes, and are schema-driven, that is, they involve top-down processing of well-learned mental templates. This may be understood when we quickly glance at an analogue watch to know the time, obtaining this merely from the shapes of the hands and their spatial relationship to the rest of the watch face. In high context cultures a less direct communication of process or task management through non-verbal communication is common; and it becomes automated over time as unconscious competencies which appear informal and heuristic. Largely they are not resonant in working memory (short-term memory). On the face of it, much of this is passed off as 'intuitive' processing, but this belies a repertoire of intricate and systematic skills drawn from a rich database of procedural (concrete scripts) and declarative (abstract) or factual information (Andersen, 1982) in long-term cultural memory.

Cultural memory systems include the organization of prior knowledge into categories. These categories are largely natural (Wittgenstein, 1953; Rosch, 1975), and may be the result of a process that is 'wired in' to abstract *prototypes* from an ever-changing universe of stimuli and data. Analogue representations are prototypes. Prototypes are schemas that evolve through learning by abstracting 'unification' rules for organizing structural sets from the world of diversity, i.e., changing environmental stimuli. The prototype abstraction process is innate, not the representations, so that greater economies in neurological storage capacity are afforded. The course of socialization in the culture then adds data to embellish these schema.

Schemas are structural prototypes that provide rudimentary 'templates' which skew the perceptions of incoming stimuli during encoding; they store the data accordingly within 'scaffolds'; and they drive a response bias. Therefore both the inherent structural form of the schema and the learned data reside in long-term memory. Long-term memory involves parallel processing and random access; it is largely unconscious. In the evolution of cultural expertise, skills are learned through distributed practice (paced with intervals) over time, with enough mass practice (rehearsal) within each interval to enter long-term memory as automated bundles.

In the real, messy world the retrieval and application of these skills would appear as if spontaneous. Retrieval cues exist in both *digital* and analogue formats. Digital codes require more conscious attention such as when we read digital watches, or tenaciously follow the syntax and decoding of the words in legal contracts or the specifications of business contracts. Conscious attention is the 'transparency' that working memory provides but affords a severely limited serial processing capacity. It slows us down. Analogue cues map more directly into analogue representations in long-term memory and involve a less conscious cognitive effort.

However, in one interesting study in cognitive psychology, it was found that even with meaningful one-word semantic cues (nouns as verbal stimuli), matching information was retrieved faster from long-term memory (Freedman and Loftus, 1971) than with cues with little meaning. Thus while not actually analogue in *representation* these verbal analogues afford, practically speaking, economical processes. As one-word *cues* are meaning-rich signals they activate 'bundles' of matching information in long-term memory. The process occurs as analogue *codes* where category size does not matter. In other words, parallel processing is implicated whether categories in long-term memory are large or small in terms of the number of component elements.

The relatively longer history of Asian civilizations directly raises the probability of larger 'bundles' of social schema being formed and being 'tripped' through parallel processing . Several thousands of years of structured civilization with little need to challenge the natural order of things allows for massive 'bundling'. Stereotyping behaviours are examples of social schemas that break down when the analogue parallel processing of matching is pushed to the limits. There is a need to *unbundle* when critical environmental features do not match prior knowledge in long-term cultural memory. You can unbundle easily in working memory but not when invoking parallel processing from long-term memory. Transparency is simply not inherent in parallel processing: yet perception here is quick (Gick and Holyoak, 1987; Larkin, Reif, Carbonell and Gugliotta, 1988).

Unbundling automatic routines is difficult. Similarly in much of relationship-building, the informal supervision in process checks that are often heuristic may be seen as the result of this automatization of cultural scripts in the course of cultural evolution. Asian culture and history *is ancient* and it characteristically integrates self with society, the environment and the natural order of the cosmos. Just as in the performance of expert car-driving where an operator appears to navigate through an environment automatically, automatic interpersonal analogue processes dominate high context interactions somewhat naturally given their longer history of navigating through relationships with the cosmos. Cultural values drive levels of pertinence to assign environmental features their due attention. If perceived as not critical they will be attended to by the process of *assimilation*. This is relatively easy with no need to uncouple from automaticity . Working memory entertains the differences in the task or relationship environment. If perceived as highly critical, then there is a need to unbundle or *accommodate*. This is of course stressful. And this is where intercultural dissonance arises.

*Transparency and Low Context*

In low context cultures communication involves 'digital' or propositional codes (verbalizations). The emphasis here is on the precision required which demands constant attention to detail. Low context (LC) societies are an artifact of the industrial revolution where rural to urban drifts of large populations have spawned discrete and smaller family units such as nucleated families. There is less formal emphasis in LC society on building relationships. Transparency is cognitive, and is part and parcel of a consciously managed process. Every layer of the process, including procedures, must be structured into steps which are clearly distinct, albeit linked. From a systems viewpoint this provides control with feedback and adaptation. And for some – the LC adherents – it offers a conscious formal management in a chaotic world: which is how they perceive the HC world with its attendant lack of transparency.

Documentation and standards are overtly and explicitly communicated in the LC world. Much of this orientation is manifested in quality management systems, performance appraisals, and competencies that break up overall expertise into a hierarchy of documented components. Appointments that specify times to the minute are expected be honoured, and punctuality is a major factor in the evaluation of someone who is late or tardy. Similarly, because of the relatively strong emphasis on verbal communication, in the LC world contracts are written and are specific. They delineate the specifications of deliverables, products, roles and time frames. Market leads have to be formally assessed with the appropriate documentation to prevent wastage of time or other resources in promotional efforts. Even contingencies are covered by legal clauses.

Consider the following case which demonstrates cross-cultural conflict. As before, names have been changed.


II


Wong Kee Seng (a Malaysian Chinese) was the country manager of a major European-based multinational enterprise resource planning software firm for 7 years. He was acclaimed as the 'country manager of the year', and the Malaysian business unit that he ran was the largest in sales revenues prior to the economic crises. During the 'miracle' years it was very easy to sell. The demand for technology, software, and technology consulting in ASEAN (Association of South East Asian Nations) seemed unstoppable. The uptake demand for a prestigious Teutonic product was of course helped by the strength of its brand name, and Wongs sales managers (on the face of it) did not seem to need managing. Of course the brand had established itself as the market leader with the best overall functionality to support the growing trend in the business process re-engineering environment of the 1990s. In 1998, Wong resigned for 'personal' reasons. This was a shock event for the employees in the firm, not to mention Wong himself.

Rumours from within the firm suggested that he had to leave 'reluctantly'. An Australian, new to the firm, was brought in as the regional Executive Vice President (EVP or Veep). The firm's regional office was in neighbouring Singapore. The new management was quick to decree that 'from now on things will be different', and that sale managers who were the front-running revenue generators and the 'stars' of the organization were not indispensable. Naturally the whole organizational culture became fearful. The main offence of Wong was that he did not manage process. He did not mandate the practice of populating the marketing information database which was apparently the standard practice of a 'best practice' global organization (Bartlett and Goshal, 1995) wherein back office standards are diffused throughout global offices.

Furthermore, the quarterly practice of creating sales forecasts and documenting the steps of enactment and progress fortnightly were not practised. Accuracy in forecasts was now demanded, and progress in the sales cycle had to be documented. How close you were to the customer had to be measured even before the closure of the deal. This was so new in terms of practice that it was a shock to the culture. In addition, the new Veep had instituted mandatory weekly sales meetings early in the week – and latecomers were vulnerable to verbal putdowns, whereas before, Wong's sales meetings were few and far between, and had an informal spontaneity which was often punctuated with laugher. The Europeans from the head office in Western Europe who were deployed in the region had revealed to the local staff that for a long time there was a desire to remove Wong. How could this happen? He delivered beyond his targets for his office, yet there was a history of dissatisfaction that had an origin beyond his performance as best country manager in the boom years. It was noted, in retrospect, that even at the time of the imminent collapse of the economy with the devaluation of the Malaysian Ringgit (by about 70 per cent against the US dollar) that the European head office was concerned with *low context* articulation. They focused on detail, even during the boom years and through the devaluation of the local currency, so pre-empting any trust developed over the terms and references of contracts by demanding that payments be made in their particular European currency.

One 'star' sales manager, who had exceeded his sales targets, explained 'We sold to the largest public corporations in the country, yet they [the Europeans at the head office] would demand that we furnish signatures of each member of the customer's board of directors as a specific revision in the terms of reference of the contract. The directors had co-signed at the bottom line for the amount of the licensing fees, but going back and chasing other directors for changes in the terms would endanger the sale. That is crazy! What is more important – bringing in the money, or shuffling paper? We took care of our leads, and managed them in our own way. It was informal but it worked.'

## Reflection

Clearly in this case there was a new thrust towards transparency, and a *managed* task orientation. More fundamentally and importantly, managing processes *formally* was perceived as a critical successful factor from the European head office's point of view. The inherent logic was simply that critical successful factors underpin key performance indicators. Sales revenue as the ultimate performance indicator was neither the end-all or be-all for the European head office. This assumption was of course not shared in the Malaysian office.

The construct of self-monitoring (Snyder, 1974), adapting to the context, modifying and controlling presentations, involves in Asian contexts a high degree of non-verbal analogue processing. In one study of Malaysian sales people, it was found that self-monitoring was positively correlated to sales performance as measured by sales earnings (Nik Mat, 1995). Although this study did not explicitly identify the non-verbal contextual messages, it is reasonable to infer that these non-verbal cues and behaviours are nested within the target population. The correlation between self-monitoring and performance earnings was significant. It appears that self-monitoring improves sales earnings in this study and that high performers indulge in more self-monitoring. Data from the US indicate that self-monitoring is unrelated to sales earnings. How could this be? Nik Mat suggests that Malaysian sales people are more adaptive, as their culture approves of personal selling where relationships are key. Adaptive selling requires heuristic approaches to identifying, interpreting and responding to contexts appropriately. Given a multicultural Malaysian business environment between the Malays, the Chinese, and the Indians (notwithstanding any outsiders), the need to be sensitive to cultural diversity is mandated in order to build relationships.

Hofstede (1980) established 'collectivism' as a dimension where cultures are either individualistic or collectivist. Malaysia, as in other East Asian nations, has a low individualism score. It is more collectivist in comparison with Western industrialized nations, which are largely individualist. Collectivist cultures place society over self, where group thinking dominates. *Face-saving* is also prevalent when confrontation and contradiction are to be avoided. They also adhere to the use of titles and salutations, emphasize relationships, and express loyalty. There is a popular consensus that high context cultures also integrate work, spiritual, and personal spheres of their lives. However there is a need to be guarded in defining the features of this interaction, as the exact sociometric relationship of both Hall's dimension of context and Hofstede's dimension of collectivism remain untested and unresearched.

Consider the same case as before with changed names.

The new Veep tells the new recruit in sales, 'When Wong was around there were in-groups. The culture was paternalistic. He [Wong] would not manage the process but seemed to tolerate a culture of coalitions that reflected his favour.' Similarly, the same high performing sales manager reports, 'In a way I'm glad that this shake-up, shape-up, or ship-out is happening. In Wong's day, requests for proposals would be faxed in, and sometimes whoever was vertically organized for that market might not receive the lead.' However, others also had this to say: 'He [Wong] was strongly liked by us. He was kind, and never breathed down our backs. He is a loving father and husband.' While there was shock emanating from Wong's departure there was a quiet unease. There was no ritual of mourning for the man who had 'fathered' the company since its formation in Malaysia seven years before. Ironically, the high power distance was now used against that previous boss who suddenly was not seen as relevant any more. Fear, with a pragmatic respect of the new Veep, inhibited an open discussion of the sudden departure of previous boss Wong.

**Reflection**

The culture while paternalistic was perceived by some in the firm as a benign dictatorship. The notable exception to the stipulated trait of formality in power distance was the relative informality permitted by the company. That informality seems to be more prevalent in regions of rapid economic development. To be sure, these models, whether Hofstede's or Hall's, maintain variations of culture as slowly changing antecedents. But culture's consequences are also evolving. Imagine that what you are witnessing in Malaysia is a time compression of all the 300 years of the industrial revolution into the 30 years of recent development. Now add to this the demands of the knowledge worker era, and the collapsing of the traditional structures' styles of governance enabled by the free flow of information. Within this demand for rapid adaptation the easiest aspect to change in power distance terms would be the formality of protocol; the most difficult to relinquish would be control. Alternatively, it could just simply be the trait of a high context culture.

Consider another case in Malaysia. As before, names have been changed.


## III


### *The Organization*

Bintang Telekom was a Malaysian company that was created from a piece of paper. It was touted as a completely radical new design of an organization that was to be enabled by a process architecture drawn from best global telco practices being enabled by high technology. Its design challenged the traditional PTTs which had been spawned by organizational design features that emphasized the functional differentiation of organizational units as a throwback of the industrial revolution era, and its logical requirement for the division of labour – where *form followed function*.

Bintang was organized along seven distinct process competencies. Instead of departments such as charging and billing, information technology, and procurement, etc., the seven core processes were aligned into a customer-focused process architecture. In other words, *form followed process*. But the process now embodied functionality in such a way that this functionality was not distinct. Its legitimacy was not mandated as bounded functional departments with the attendant artifacts of turf, bureaucracy but with 'unbundled' isolation at worst, or loose coupling with the rest of the organization at best. The process architecture was designed to provide seamless customer responsiveness and involved ongoing interdisciplinary deployment (as opposed to temporary cross-functional matrix teams within functional organizations).


### *The People*

The new CEO was a Malay who was a *Melayu Baru* which means 'New Malay'. He was characterized by a high need to achieve (McClelland, 1961) balanced with the spiritual and social goals of the community. Being *Melayu Baru* meant showing a high task orientation, being progressive and successful yet preserving the best traditional values of collectivism. It was the *zeitgeist* of the times in modern post-colonial Malaysia. The government was all for it.

Malaysia is a multiracial society of Malays, Chinese and Indians who respectively make up 70 per cent, 20 per cent and 9 per cent of the population. The majority group, the Malays, who are officially dubbed 'sons of the soil', or *'bumiputra'* in Sanskrit, had found themselves to be lagging behind the more aggressive overseas Chinese Malaysians who dominated the commercial landscape. After the racial riots of 1969, Malaysia had instituted a New Economic Policy (NEP) which was a programme of affirmative action to help the Malays in education, administration and commerce; it was to balance the distribution of wealth, power and upward mobility.

## The Background National Culture

Within this context of frustration and aspiration, collectivism in the Malay culture had legitimized a need for high-achieving individuals to break the mould yet be perceived as communal role models. It was inevitably destined to have a direct impact at a personal level. The new CEO had a fraternal relationship with the conglomerate's board of directors that had created the telco start-up. This relationship had its roots in a famous Malay school which had groomed many of Malaysia's post-independence leaders. Their roots and bonds invoke a sense of high context that is thick and strong: even so, the CEO was a sharp dresser in expensive Italian suits, and carried a direct off-the-cuff style that seemed incongruent with the traditional Malay warmth, and respect for others.

## The Inevitable Cross-Cultural Encounters

Bintang Telekom had also brought in a large global consulting firm to help its systems integration, change management, and strategy consulting. It recruited expatriates in the telco industry from Australia and the West. At one point the Bintang environment had 200 of these international expatriates who were crusty veterans in the telco industry. Being brand new, the organization demanded people with skills to enable its cutting-edge competency. Personnel, as raw material, in the tight labour market of Malaysia were hard to find. Strategic HR had identified that the management of change was vital even if the company was new. What had to be done was the creation of a customer focus culture largely drawn from the Malay talent pool. And Western expatriates had quickly noticed that there was a need to empower the workforce. One of the Westerners in customer care (originally from Australia), having several years of telco experience says, 'These young recruits seem bright, but need to be empowered.' A formal culture audit was then commissioned by the consulting firm which identified that the senior management needed to articulate a strong, central vision to integrate diversity, and manage dissent.

On the other hand, diversity was necessary as the telco had contracted Western experts for a period of two years to support skills transfer to locals. Further, it was clear that a long-term mindset provides the context to interpret daily events, provide meaning, and to give motivation at all levels. In other words, strategic direction provides confidence in organizational continuity. It motivates. This is an especially important point since these were the boom years and many businesses were being formed overnight. As one senior British expatriate in the consulting firm had put it, 'We are dealing with "knee-jerk" Malay entrepreneurs. . . . but the process of creating a new culture has to be managed formally not informally. And the process of changing a workforce culture has brought to the workplace a national cultural baggage. Can it be done? Yes.'

**Reflection**

*Input of Economic Drivers on Collectivism and Power Distance*

On the face of it this 'knee-jerk' view may seem like a 'cheap shot' indictment against Malay and Asian entrepreneurialism in general. But consider the context. Malaysia was cash rich in the mid-1990s. Its economy had averaged a real GDP growth rate of 8 per cent over the previous eight years. Further, their GDP had been in the double-digit zone at least twice. The mood was euphoric, and optimism was high. Any call for discipline and managed growth in the economy was immediately perceived as party-pooper talk. Growth seemed unstoppable. Cultural theories provided insights that helped identify opportunities, and obstacles, in a high-growth environment. Recognition of the collectivism, power distance and high context indices had forewarned and forearmed the culture creation enactment. It was known that the inherent high power distance, collectivism and context presented an opportunity to deploy the said CEO as a sponsor to initiate change. His authority was already formally designated, and the skills transfer was targeted to a largely Malay culture that was warm and that enjoyed relationships and ritual. It was easier to foster a coherent culture that was collectivist, that respected and followed formal authority, and that naturally valued relationships and the process of building them. But there were complications.

*Impact of Collectivism on Individuality*

The CEO wanted to create a task-oriented culture and did not want a national East Asian culture of collectivism and high context to be an excuse with its relatively strong emphasis on relationships over task. He was perceived also to show little empathy for his employees. He did, however, want a climate of fun and informality 'to take the heat off' an organization that was ramping up very rapidly for new operations. There was a largely invisible pressure on him to look like he meant business, to be entrepreneurial. And he had to guard the financing of the company by securing large loans, as Bintang was a commercial enterprise, and financially independent of the state.

The government, through their New Economic Policy (NEP), have generally empowered the bumiputra. The government also intimated that it was the responsibility of these people to deliver the policy and it was expected that those who were so empowered would be role models for the community. It was in this context that the CEO had inherited this firm, even though his business was independent from the state. As in any business there was the pressure to prove the profitability with all the investments made. But for him, cultural sense making and tolerance had to be weighed against intervening factors such as the stages of national development and the national agenda to create rugged Malay entrepreneurs. He had to ride and drive the rapid economic growth and the pressures in all start-up new industries. Core competency development, and their inherent evolutionary paths, were seen as a luxury in a rapidly developing Malaysia. Management was expected to be flexible and responsive to accommodate new competencies. One further difficulty was that Bintang, as a new telco, belonged to a property conglomerate which like other property, plantation, and extraction conglomerates wanted to diversify.

## Role of Government in Collectivism and Power Distance

Furthermore, Bintang's parent conglomerate had been instigated by the government, and the present government officials still had strong relationships with the conglomerate's board. This was very much part and parcel of the NEP. It was a perfect case of collectivism with state blessings, and had origins in the traditional pre-industrial ethos of feudalism. Is this wrong? From a collectivist, high context, high power distance, Asian point of view – and more specifically a Malaysian one – the answer is no: it is not wrong. Governments can and should be involved in the creation and promotion of new industries, to facilitate the migration of skills from commodity-based industries to knowledge worker industries. Government has the moral and legitimate, not just legal mandate for a national agenda, even though it may be branded as 'interventionist'. This is indeed an evolutionary leapfrog in economic development encompassing social, cultural and psychological integration, allowing it to ambitiously and aggressively by-pass an industrial manufacturing era. Government is the agency that represents coordination and control in collectivist societies. It therefore has a high context and natural right to oversee and facilitate.

### Why Control is More Difficult to Relinquish than Formality

Given that structures of governance and control mechanisms could well be centrally internalized over the long Asian history of the evolution of cultural schemas, there may therefore be deep structures that are difficult to unbundle. Uncoupling these structures may indeed be stressful, and anxiety-ridden, in high power distance and collectivist cultures. By that same token it can be argued that the need for formal protocols in high power distance, collectivism or high context cultures can also be just as ancient, centrally stored – and be just as stubborn. So why should they change more easily if control remains more difficult to relinquish? The answer could be that control carries more perceived valence and costs than formality. In the extreme pressure of time compression there is a need to adapt from feudal systems to task-oriented organization architectures that will support formality, but not suffer from the loss of control. This proposition of course remains to be tested through the rigours of social cognition research. At an individual level it seems easier to look informal while feeling the need to hold on to control.

### Managing the Context of the Learning Organization

The culture audit of Bintang revealed some inherent antagonisms. Collectivism, high power distance and context implies a desire to preserve the status quo and, indirectly, longevity (Abdullah and Gallagher, 1995). The need to balance strategic long-term vision with the urgency of daily operational tasks was necessary. If this was not achieved, roles and activities would *not* be integrated into a meaningful, stable context of an organization that desired to emerge as a long-term focused world-class telco. It could not afford to be perceived by employees as a short-term telco *vis-à-vis* other national telcos; especially *if* a project were to flounder in a market that had a high labour turnover. There would be no skills transfer because there simply would be no motivation to learn without context predicated on a perceived long-term viability.

### Postscript

Interestingly, the mental migration from a functional department based organizational architecture, often dubbed as 'silos', to a process architecture was also to prove difficult, even for the experienced Western expatriates. They had come from telcos that were departmentally organized into functions. If for these veterans the process architecture was too overwhelming, with the attendant demands of multi-tasking across different competencies, what could be said for the new local recruits? The customer-focused process architecture was the glue that was to bind different sub-processes together to provide the efficiencies of seamless integration with shortened cycle times. Now it was in jeopardy. It demanded perfect knowledge based upon perfect transparency. True, there were electronic on-line performance support systems in place that were high in useability, and provided 'just-in-time' information on site and on demand to enable this process architecture. But there was no transition in task enactment to facilitate transfer of prior experience for the 'experts'. They were experienced people, and were recruited to 'hit the decks running'. They found these to be certainly unfamiliar decks both on cultural and task fronts.

As an intermediate solution, the functionality was restored to the organization to allow a learning curve to take place. Low context cultures carry a trait dubbed *monochronicity*, i.e., time is perceived as compartmentalized rigidly to encapsulate schedules. There is a sequential unfolding of task and communications which are dealt with in a linear mode. Customers are accordingly served one at a time, one after the other. Schedules and time-tables are tenaciously adhered to so that deadlines are met. Planning favours algorithms. Low context cultures emphasize tasks and they compartmentalize relationships which mandate a tight agenda in the acquisition and use of information from various sources. Relationships are organized around tasks.

High context cultures are said to be polychronic. The defining feature is accepting a simultaneity of events without closure. Time here is viewed flexibly. Enactments may be heuristic, completion of communication and tasks are considered more important than following a schedule rigidly. Polychronic time is said to be cyclical and multitrack: and it is perfectly acceptable in high context cultures to carry on multiple tasks and engage in more than one conversation simultaneously. Strangely people appear less hurried and seem to have less stress. There is a higher propensity to appear accommodating, and information flows through fluid links in a broad network of interpersonal contacts.

It appears that some of the elements of polychronicity may be exploited and synergized over time to accommodate a higher degree of task orientation. The development of a process architecture might have exploited this polychronic style, and over time the local culture could have been trained to attend to the tasks in the appropriate customer focus sequence. Unfortunately, the project to migrate to a complete process architecture was disrupted by many reasons including the economic crisis. The level of unbundling for the high context orientations to sequential customer service and planning would of course have required more time.

## CONCLUSIONS

The management of process has been discussed firstly on (micro) cognitive levels where transparency may be seen as a cultural artifact of the conscious thinking capacity of the mind in low context cultures. This capacity is described by cognitive psychologists as working memory – which has a limited capacity and processes information in serial or sequential modes. Long-term memory is unconscious, involves schema-based parallel processing, and is the repository of cultural memory in high context and ancient East Asian cultures. Social schemas are cognitive templates in long-term memory that shape perceptions, drive expectations as silent 'action scripts' and hypotheses (Fiske and Taylor, 1991). By and large they give permission to the popular currency of stereotypes be they ethnic, occupational or gender. These social schemas also slowly evolve, and are very resistant to change. They represent the more central structures and processes in human cognition and subsume cultural schemas. High context cultural cognition is largely shaped automatically and the management of process is unconscious and informal.

Secondly, the management of process has been shown to require formalization in the task environment demanding a sensitivity towards unbundling the automatic cultural schemas of East Asian cultures. Transparency, in the conventional sense of the phrase, is a formal delineation and explication of process steps into procedures with formal documentation. It also includes the specification of contracts, punctuality, management by objectives, individual accountability, and so on, which are presented as 'digital codes' in low context cultures characterized by Western business practices. There is a tremendous volume of 'unbundling' required of nested relationships within the older social order, in the cultural schemas, and there is the need to create task-oriented cultures that focus overnight on acquiring formal process management skills, attending to low context specificity, and documenting and maintaining conscious vigilance. Further, in creating knowledge workers, there is a need to transform data through information, knowledge management, and business wisdom within the cognitive chain of different levels of processing (Craik and Lockhart, 1972).

The practical application of this value-adding chain of information literacy in the management of the cultural environment demands not only empowerment, but managed responsibility. In the management of cognitive development, 'mindfulness' (Langer, 1997) and the sustenance of mindfulness (Senge, 1999) support mechanisms must be in place both formally and informally as infrastructure. Unbundling control mechanisms seems harder. Schema research reveals that schemas are slow to evolve and once established are relatively stubborn.

For example, a popular quality consulting viewpoint and lament in East Asia is that ISO 9000 and total quality management systems require proactive transparent, system-wide coherent organizational support. Quality has to be built into a transparent and formally managed process within the system. Chief executives also need to sponsor and drive the quality initiative as a process and not an event. Limited research in Malaysia has indicated that there is little relation between ISO certification, TQM initiatives and profitability (Jabnoun, 2000; Meek *et al.*, 1997); and that quality initiatives may be undertaken for many reasons, even as promotional exercises.

While high context cultures are naturally coherent they are informal in communications. Further, high power distance may inhibit participative management that solicits inputs from workers by their formal hierarchies. Various stages of industrialization have to be assimilated into this model of high context, collectivism and power distance so that integration of informal social coherence into teamwork, and formal hierarchy into legitimate sponsorship, can be exploited in quality and change management. For instance, Singapore, as a city-state, ranks first with the USA in competitiveness as measured by the International Management Development (IMD), and represents a country with strong social re-engineering. Here transparency is mandated, instituted and formally managed from 'outside' the individual by the state rather than leaving it to the spontaneity of Asian culture.

Even today, all ASEAN cultures represent high context, collectivism and high power distance. They have similarities but with critical differences. There are of course inherent weaknesses due to over-aggregating Hofstede's data (Sondergaard, 1994), and generalizing beyond sample constraints (Lim, 1998). The differences amongst Malaysian Chinese and Malays were not examined specifically in this chapter: they are summarized by Lim (1998) as being more inclined towards affiliation needs, i.e., lower in masculinity (Hofstede, 1991a), having stability, and honouring tradition. Malaysian Chinese are said to be more materialistic, higher in risk taking, strongly persevering and thrifty (Hofstede and Bond, 1988).

**New Directions**

Another perspective, that of Vedic management, fruitfully combining the best of Western empirical rigour and Eastern subjectivity, is offered here as a universal management model that fulfils the need to adapt and preserve cultural integrity. This view allows a fundamental insight of how consciousness underpins the cognitive act of interpretation. Western models of management may be argued to be predicated upon a Newtonian metaphor of localization of action and reaction, discrete mechanics, and point-value communication largely based upon the limited capacity of the conscious mind and within the working memory.

In Vedic and Eastern 'management' there has been an acquiescent integration of self with society, and society found no need to formally manage given that no 'professional' models of management have been developed. This is the largest and deepest value of 'high context' that allows for 'textures' of variety. In traditional Eastern culture, 'management' was taken for granted as an unseen, silent default. The roots of this thinking can be traced back to Taoist and Vedic philosophies of management within the unified field of Nature. However there obviously was, and is, a breakdown of this Eastern mode with the onslaught of the industrial revolution from the West bringing with it its baggage of alienation, the creation of new classes, and thus class conflict. Integration in Eastern cultures was a 'given', grounded in communal ties and relationships, which was and still is part of the quantum mechanical texture of creation where individuals, families, communities, government and enterprise were interwoven and invisibly connected. Communication was and is also non-verbal in a high context of stimuli.

Thus thinking and feeling were seen as compatible where feeling can even be basic, and drive thinking such as intuition. Western psychology, in contrast, sees affect as being for the most part post-cognitive, and it demotes affect to be subservient to rational thought. In Eastern psychology, affect is invisible and it is intuitive. It is understood by the unconscious mind as refined feeling levels of experience. Notable exceptions to this established paradigm have already taken hold in Western psychology as seen in Zajonc's (1980) seminal paper 'Thinking and Feeling: Preferences Need No Inferences'. Affect is postulated to be independent of thinking and indeed even basic to thinking in the exercise of judgment.

We face the post-industrial age of knowledge and wisdom with societies enabled by new technologies in electronic commerce and communication. New intelligence driving new competencies is required, where time and space are shrunk, if not collapsed. To overcome the negative transfer effects of learning new cultures, 'mindful' (Langer, 1997) unbundling has been suggested by cultural trainers to consciously understand the bases of schema and values that drive our orientations. But it may be more practical and natural to expand the conscious capacity of the mind by meditation methods – Transcendental Meditation or TM for example. TM is also grounded upon the principles of a unified field. Vast research in TM has shown the spontaneous development of Western operational definitions of intelligence in cognitive and perceptual tasks, as well as wider dimensions of intelligence and adaptability, improved social behaviour and field independence. For example, negative transfer was found to be reduced by TM meditators, exhibiting little interference from prior knowledge schemas. (For another view of TM, see Heaton and Harung in this volume – eds.)

There is no single approach, superior to the rest, by which the East Asian may develop. There is a need in the 'East' to adapt from the 'Western' models of best practices with their transparencies in organizations with a large workforce and complex activities. There is also the need to develop the intelligence and wisdom of the 'knower' drawn from 'Eastern' technologies, and to further recognize culturally valid, albeit 'invisible, fields', so harnessing these where appropriate.

# References


Abdullah, A. and Gallagher, E. (1995) 'Managing with Cultural Differences', *Malaysian Management Review*, 30(2), pp. 1–18.

Alexander, C. N. (1993) 'Transcendental Meditation', in R. J. Corsini (ed.), *Encyclopedia of Psychology*, 2nd edn (New York: Wiley Interscience).

Andersen, J. R. (1982) 'Acquisition of cognitive skill', *Psychological Review*, 89, pp. 369–406.

Bartlett, C. A. and Ghoshal, S. (1989) *Managing Across Borders: The transnational solution* (London: Century Business).

Berry, J. W., Poortinga, Y. H., Segall, M. H. and Dasen, P. R. (1992) *Cross-cultural Psychology: Research and Applications* (Cambridge: Cambridge University Press).

Craik, F. and Lockhart, R. (1972) 'Levels of Processing: A Framework for Memory Research', *Journal of Verbal Learning and Verbal Behavior*, 11, pp. 671–84.

Hagelin, J. and Herriott, S. (1991) 'Unified Field Based Economics', *Modern Science and Vedic Science*, 4, pp. 72–95.

Fiske, S. T. and Taylor, S. E. (1991) 'Self-Efficacy and Personal Control', in S. T. Fiske and S. E. Taylor (eds), *Social Cognition* (New York: McGraw-Hill) pp. 197–204.

Freedman, J. L. and Loftus, E. F. (1971) 'Retrieval of Words from Long-Term Memory', *Journal of Verbal Learning and Verbal Behavior*, 10, pp. 107–15.

Gick, M. L. and Holyoak, K. J. (1987) 'The Cognitive Basis of Knowledge Transfer', in S. M. Cormier and J. D. Hagman (eds), *Transfer of Learning: Contemporary Research and Applications* (San Diego: Academic Press) pp. 9–47.

Hall, E. T. (1976) *Beyond Culture* (New York: Anchor Books).

Hofstede, G. (1980) *Culture's Consequences: International Differences in Work-Related Values* (Beverly Hills: Sage).

Hofstede, G. (1991a) 'Management in a Multicultural Society', *Malaysian Management Review*, 26(1), pp. 3–12.

Hofstede, G. (1991b) *Cultures and Organisations: Software of the Mind* (Basingstoke: McGraw-Hill).

Hofstede, G. and Bond, M. H. (1988) 'The Confucius Connection: From Cultural Roots to Economic Growth', *Organizational Dynamics*, 16, pp. 5–21.

Jabnoun, N. (2000) 'The Impact of ISO 9000 Certification on Financial Performance of Service and Manufacturing Companies', presentation to the 12th Annual

Asian-Pacific Conference *International Accounting Issues*, 21st October, Beijing, China.

Langer, Ellen J. (1997) *The Power of Mindful Learning* (New York: Addison-Wesley).

Larkin, J., Reif, F., Carbonnell, J. and Gugliotta, A. (1988) 'FERMI: A Flexible Expert Reasoner with Multi-Domain Inferencing', *Cognitive Science*, 12, pp. 101–38.

Lim, Y. Lrong (1998) 'Cultural Attributes of Malays and Malaysian Chinese: Implications For Research and Practice', *Malaysian Management Review*, 33(2), pp. 81–8.

McClelland, D. C. (1961) *The Achieving Society* (New Jersey: Van Nostrand).

Meek, G. E., Sudin, b. Haron, Lim Kong Teohg and Hasbulah b. Hj. Ashari (1997) 'The Awareness and Usage of TQM Tools and Concepts in Small Businesses in Malaysia', in M. Sulaima, D. N. Ibraham, Z. A. Ahmad and Mohd. Shaffie Ariffin (eds), *Towards Management Excellence in the 21st Century Asia*, Proceedings of the 2nd Asian Academy of Management, Langkawi, Malaysia, 12–13 December, pp. 280–3.

Nik Mat, Nik Kamariah (1995) 'Personality Factors for Successful Sales Performance in Malaysia', *Malaysian Management Review*, 30(4), pp. 32–54.

Rosch, E. (1975) 'Cognitive Representations of Semantic Categories', *Journal of Experimental Psychology: General*, 104, pp. 192–253.

Senge, P., George, R. and Bryan, S. (1999) *The Dance of Change: The Challenges of Sustaining Momentum in Learning Organizations* (New York: Doubleday).

Snyder, M. (1974) 'Self-Monitoring of Expressive Behaviour', *Journal of Personality and Social Psychology*, 30, pp. 526–37.

Sondergaard, M. (1994) 'Hofstede's Consequences: A Study of Reviews, Citations, and Replications', *Organization Studies*, 15, pp. 447–56.

Wittgenstein, L. (1953) *Philosophical investigations* (New York: Macmillan).

Zajonc, R. (1980) 'Thinking and Feeling: Preferences Need No Inferences', *American Psychologist*, 35, pp. 151–75.

# Part III

# Making Sense of the 'Us and Them' Scenario

# 9 Beyond Culture Shock: Developing Cross-Cultural Awareness

Lionel F. Stapley

## INTRODUCTION

The aim of this chapter is to provide the reader with a way of thinking about the effects of societal culture on our behaviour and about developing cross-cultural awareness that will result in more productive working relationships. I shall start with a clarification of what is meant by culture using my previously developed work referred to in *The Personality of the Organisation: A Psychodynamic Explanation of Culture and Change* (Stapley, 1996) and use this to provide an explanation for the way that diverse cultures develop. I shall then explore the notions of 'organizational socialization' and 'culture shock' as a means of explaining what happens when we operate in an alien societal culture. This will lead, in the latter part of the chapter, to the development of a way of achieving cross-cultural awareness and understanding that will create effective relationships.

I am taking as a given that there are a multitude of different cultures and sub-cultures, both societal and organizational, that managers of multi-national companies have to contend with if they wish to do business on a global basis. I shall not, therefore, relate to any particular culture but will provide a general theory that can be applied to any society or organization. However, I would point out that the reader will only gain by applying the approaches developed in this chapter to their own experience. By self-reflection you will gain a greater self-awareness and through this process be able to gain an understanding of others. Whatever the society or culture, it is your experience that matters, not my interpretation of it. At best, I can only provide a much condensed and sanitized version; at worst, I might provide a totally inaccurate version.

## WHAT IS SOCIETAL CULTURE?

The term 'culture' has been in common usage for many years and is a familiar notion. Unfortunately, this common usage has itself led to problems. For example, cultural differences are seen as being in the nature of things requiring no explanation. A result is that functions which are not easily understood are assigned to a mysterious central agency called 'culture' accompanied by a declaration that 'it' performs in a particular way. Culture is also an easy option to fall back on to solve all our unexplained problems. In addition, past uses of the word to designate a way of life such as a particular society, or part of a society, are exceedingly vague. However, as vague as our understanding of culture may be, it is also vital to our understanding of different societies and in particular to our understanding of cross-cultural relationships.

Societal culture is a complicated yet highly influential phenomenon and the degree of complication is likely to be much greater in a situation where multiple societal cultures prevail.

I would suggest that in the past we have been more concerned with identifying the nature or the symptoms of a culture rather than understanding what it is. My approach seeks to provide an answer to that most fundamental of questions, that concerning how culture develops, working from the principle that if we know how it develops we shall be able to unpack it and therefore know how to influence it. In other words, knowing how culture develops will provide us with an understanding of the causes of the consistent behaviour that we call culture.

So, what is it that we are we trying to understand? In simple terms, we are trying to understand processes of human behaviour, which by their very nature are dynamic, that is they exist in a state of flux and are characterized by spontaneity, freedom, experience, conflict and movement. So how do we study these dynamic processes? What is absolutely clear is that the value-free, value-neutral, value-avoiding model of science that was inherited from physics, chemistry and astronomy where it was necessary to keep the data clean, is quite unsuitable for the scientific study of life. When we extend science to the extreme difficulties of biological behaviour, human emotion and social organization then other sources of insight and methodologies beyond those of the empirical analytical scientific method must be sought.

Bhaskar (1975) pointed out that the social scientist can only say as much as the tools at his disposal, or those which he chooses to use. The tools used here to provide the sought-after explanations are concepts from psychoanalysis which I believe provide explanations for human behaviour that are both consistent and comparable. I will gladly acknowledge that working in this way is not a perfect art. However, by adopting this approach we can discover enough of the underlying dynamics of the total situation to enable us to gain an awareness of societal cultures.

The constant interaction between the individual and culture is fundamental to any study of culture, or for that matter, personality. They are indivisibly linked and consequently it will be necessary to refer to both processes. Indeed, we may take as our first premise the notion that the function of the personality as a whole is to enable the individual to produce forms of behaviour that they feel will be advantageous to them under the conditions that they perceive are imposed on them by their environment.

The influences which culture exerts on the developing personality are of two quite different sorts. On the one hand, we have those influences that derive from the culturally patterned behaviour of individuals towards the child. These begin to operate from the moment of birth, a matter which will shortly be dealt with in greater detail. On the other hand, we have those influences which derive from the individual's observation of, or instruction in, the patterns of behaviour characteristic of their society. The fact that personality norms differ in different societies can be explained on the basis of the different experiences which the members of such societies acquire from contact with those societies. However, what we need to study is the processes of these societal experiences.

Every society consists of individuals developing from children into parents. In the earliest days the mother provides the context in which development takes place, and from the point of view of the newborn she is part of the self. She provides a true psycho-social context: she is both 'psycho' and 'social' depending on whose perspective we take, and the transformation by which she becomes for the infant gradually less 'psycho' and more 'social' describes the very evolution of meaning itself. In Winnicott's (1971) view, what he refers to as the 'holding environment' is vital to the development of the infant. From the beginning of life, reliable holding has to be a feature of the environment if the child is to survive.


## THE MATERNAL HOLDING ENVIRONMENT

The notion of a 'holding environment' is seen as the key concept in providing an explanation of how culture develops. Holding in the mother's womb and then holding in the mother's arms, is the first boundary out of chaos within which the infant's personality can develop. The early relation in the maternal holding environment is characterized by infantile dependence, that is, a dependence based on a primary identification with the object, and an inability to differentiate and adapt. A relationship grows through the ability of both parties to experience and adjust to each other's natures. The relationship develops through the infant getting to know the mother as she presents herself to interpret and meet his needs, which are emotional as well as physical. For the infant to develop there is a need for a 'basic trust' in the maternal holding environment and for what Winnicott (1971) has referred to as 'a good enough holding environment'.

As the infant grows there develops the formation of a self-concept. This psychological change arises once the infant is able to experience the mother and other significant objects as separate. Gradually there develop several 'not me's' in the shape of father, siblings, playmates and other relations. At this stage the infant is capable of introjecting cognitive symbols. And, here, the holding environment begins to split into an *internalized psychological part* and an *external social part*. By 'taking in' (introjecting), 'summoning up' and 'holding in mind' their perceptions as if they were an object, infants feel that they contain within themselves a world of concrete things of at least as much reality as the material world.

Early introjections (taking in of external objects), since they are virtually all the infant has, are particularly potent, and the inner 'objects' (the mental images) they create are never forgotten. These early introjections, which of necessity are of parents or parental figures, create an inner object commonly referred to as the conscience or in technical terms the superego. The introjection of the 'good' parent creates what I shall refer to as the ideal conscience, that is, a sense of ideals and positive morality – a pattern of what to do. And introjection of the 'bad' parent creates what I shall refer to as the persecutory conscience, a sense of guilt and negative morality – of what not to do. Conscience, then, is built up by identifying with, that is forming and taking in and retaining, mental images of parental figures. It will be realized that these images may not be built on the reality of parents' behaviour but on the way that the infant perceives reality, which of course may be total fantasy. Thus the reality may have been that the parent was very caring but his or her behaviour was perceived as uncaring by the infant. Introjection is a very important concept, which simply involves the creation of an internal object which may be another person, a quality of another person, or a concept, and may include family and societal values.

To summarize, the 'maternal holding environment' consists first of the mother and child and later the father and other important relatives. In this 'holding environment' there is a continuing interrelationship between the mother and the child. The mother influences the child and the child influences the mother. In other words the child is part of the 'holding environment' and influences it while at the same time the child is influenced by the 'holding environment'. The maternal holding environment is not a closed system but is open to external influences such as, for example, noise. The development of the personality of the child will depend upon whether the holding environment has been 'good enough'.

Building on the concept of a 'maternal holding environment', the growing infant becomes a member of several holding environments: the family, the school, the university, the organisational or work, and, the societal holding environments. Indeed, I will go further than this because I believe it is more accurate to state that there is not only a succession of 'holding environments' but that several 'holding environments' may be available for any one individual at any given time. This may be especially so regarding societal, work and family holding environments which we will all influence and at the same time be influenced by.

## ORGANIZATIONAL HOLDING ENVIRONMENTS

Having demonstrated the importance of the maternal holding environment in the process of personality development, I shall now develop the concept of an 'organizational holding environment' before moving to societal culture. It is my contention that in a similar manner to the relationship of the individual with his maternal holding environment, the organization also becomes a *partly conscious and partly unconscious holding environment* for its members.

As part of the organizational holding environment – and in common with the maternal holding environment – we each influence the organizations we are in and they influence us and our behaviour. But, of course, the reader will appreciate that, unlike the maternal holding environment, there is no 'mother' in the organization setting. Basically, what happens is that we identify with the organization 'as if' it were real. It is what we might refer to as an 'organization held in the mind'. It is a construct that we identify with and treat 'as if' it were real.

In much the same way that we interrelate with the maternal holding environment so we interrelate with the organization holding environment. We use it to supply the same needs as the maternal holding environment and we apply the same affect to it and create similar defences when it is seen as 'not good enough'. There is never total independence, the healthy individual does not become isolated, but continues to be related to the environment in such a way that the individual and the environment can be said to be interdependent. In much the same way as there is a need for a 'basic trust' in the maternal holding environment and for what Winnicott termed a 'good enough holding environment' if the infant is to develop, so there is a similar need here in the organization holding environment.

As with the mother, such 'basic trust' is developed as a result of the perceived experience of the organization holding environment by the members of the organization.

In the same way that unconscious forces operate in the maternal holding environment so they are at play here in the organizational holding environment. Consequently, it is not only helpful but also necessary to view the organizational holding environment as consisting of two parts. In this respect, the 'iceberg' analogy previously used by other writers may be a useful way of viewing things. The physical or sociological part of the holding environment is that part which is exposed or is conscious. This is *the external holding environment*, which consists of: the formal structures and strategies, the ruling coalition or leader, the organizational tasks – that is, the reason why the organization exists – the roles of the various members, and all forms of knowledge and skills, values and attitudes shared by the members. The psychological part of the holding environment is internalized and largely unconscious. This is *the internalized holding environment*, which consists of the internal objects which are regarded as part of the self and compose the basic social character of the individual.

As stated earlier, culture develops out of the interrelatedness of the members of the organization and the organizational holding environment. The organizational holding environment consists of the totality of the organization including the members of the organization themselves. The specific organization holding environment provides the context in which development of culture takes place. However, there are various aspects within that totality that will have a particular influence on the members' perceived notion of the organization. In the maternal holding environment, the particular influence was the mother and later the father and other important relatives. In the organizational holding environment it may be the chief executive, the ruling coalition or other significant figures; in others it may be the strategy (or lack of), or structure.

The way that the members of the organization perceive the 'organization in the mind' will determine the culture. It will depend whether or not there is a 'basic trust' and whether the holding environment is viewed as 'good enough'. The behaviour adopted by the members of the organization will depend upon their psychological perception of the organization holding environment, or, their view of the 'organization in their minds'. Whatever their view, members of the organization will adopt forms of behaviour that they feel are appropriate to them under the circumstances that they perceive are imposed upon them by their holding environment. The end result is consistent forms of behaviour that have previously been defined as 'the way things are done around here'.

## SOCIETAL CULTURE

As has been demonstrated, the issue of organizational culture is complicated enough and we can start from the assumption that societal culture is somewhat more complicated as there are even more influences than in the organizational setting. Individuals, as Freud so shrewdly observed, belong to many groups, have identifications in many directions and a variety of models upon which to build an ego ideal. Or, to put it as was stated earlier, we become members of several holding environments.

Complicated as it may be, by building on the approach outlined we can begin to make sense of societal culture. As with organizations, we can say that societal culture develops out of the interrelatedness of the members of society and the societal holding environment. Starting from the notion of a *societal holding environment* we can begin to unpack societal culture and to establish how it has developed. Again, as with organizations, we can view the societal holding environment as consisting of an *internalized psychological part* and an *external social and physical part*.

*The external holding environment* will consist, among other things, of significant leaders such as the Prime Minister and Queen in Britain; the government and the policies and structures that they adopt; local government and the policies and structures that they adopt; and the media. In the most general of terms, these are likely to be characterized by a degree of formality and conservatism with personal space and independence being valued. In the USA, one of the significant leaders in the external holding environment will be the President; other influences will be much the same as in Britain. Again, in general terms, these are likely to be characterized by informality, directness, competitiveness and an achievement orientation. The experience of the Asian external holding environments will also include significant leaders and their policies. And, in general terms, these are likely to be characterized by formality, courtesy, modesty, humility, and collective approaches.

*The internal or psychological holding environment*, which, you will recall, is largely unconscious, consists of internal objects which are regarded as part of the self and compose the basic social character of the individual. These are derived from life experience: objects that have been previously introjected. These will include the mental images that we created of our parents or parental figures; a sense of ideals, values and positive morality that represent a pattern of what to do; and those which represent a sense of guilt and negative morality – of what not to do. These mental images, whether positive or negative, may include another person such as a political leader or mother, a quality of another person or a concept. Clearly, these will differ from society to society. However, by a process of self-reflection it will be possible to develop an awareness of your own specific holding environment.

From the perceived view of the societal holding environment, members of society develop a construct of the 'society in the mind'. We are all members of several holding environments, and we are also social animals who know that our existence is absolutely tied to the group. As members of society we test reality by comparing our own perceptions and evaluations with those of other persons who experience the same or similar events. The 'truth' of reality concepts is not tested in the isolation of lonely contemplation but in the bubbling cauldron of consensus. I would suggest that the most influential of these groups are the family and work; or to put it another way, the family and work (or business) holding environments. This is where we spend the vast majority of our daily lives and this is where we most frequently test reality with others.

We experience the societal holding environment through our perceptive processes and filter them down and match them against the pool of internalized information consisting of our past experience. The various identities that we take will affect our perceptive filters as will the holding environments that we are a part of. Therefore, based on current and past experience, and conscious and unconscious processes, the internalized multiple experiences result in a construct that we may refer to as 'the society in the mind'. It is this construct of a society in the mind that members of society interrelate with. This leads to the way that societal culture develops, because, having developed this construct of the 'society in the mind' the members of that society then adopt forms of behaviour that they feel are appropriate to them under the circumstances that they perceive are imposed upon them by their societal holding environment. The resultant behaviour is the societal culture.

As a means of demonstrating how the various holding environments affect the societal culture, the following paragraphs provide a somewhat condensed application to Chinese society. I would remind the reader, though, that you will only gain the true benefits of this chapter if you apply the approaches to your own experience. Whatever part of the world you are from, by self-reflection you will gain a greater self-awareness and through this process be able to gain an understanding of others.


## AN APPLICATION

In Chinese society the maternal and family holding environments set the scene with internalized values of loyalty, respect for the authority of parents and elders, and a stress on harmony and limitation of conflict. In the family holding environment everyone is expected to work for the common good of the family. They are expected to compromise and be willing to sacrifice self-interest for the achievement of harmony and long-term relationships. All childhood development is aimed at a common goal. Respect for authority is an important value and it means that Chinese children listen to what the parents want. One of the results of this emphasis on respect is that all decision-making is parental, and children are permitted no discretion, being expected to follow the rules.

The influence of the family holding environment is extended into the organizational holding environment where family values tend to dominate and influence. Many organizations are family firms where the owners control the firms, which are passed on from generation to generation, usually to the eldest son who is expected to respect and obey the father. Here, the same sort of family values are in operation, a philosophy based on the notion of the family and working like a team. Employees are willing to compromise and willing to sacrifice themselves for the achievement of harmony and long-term relationships. Employees feel obliged to the company, and family values mean that they are working for the company out of loyalty. This works both ways because it is not easy to dismiss anybody even if they are not a good worker. The belief is that more leniency will result in a building of trust which will lead to greater loyalty.

At the societal level, the experience of the maternal and family holding environments and later the organizational holding environments will influence the culture. However, other factors will also influence and have an impact on the societal holding environment. Paramount among these are the formal policies, strategies, structures and style of management of: central government, in the shape of the President and Chinese Communist Party (CCP) Chairman; the Prime Minister; the National People's Congress (NPC); and, the Politburo; local government, in the shape of the local people's governments in the Provinces, Autonomous Regions and Municipalities; and the CCP. The latter, with its hierarchical committees functioning from village level upwards and taking orders from above, serves to emphasize and support the family values referred to above.

As is the universal case, Chinese people adopt forms of behaviour that they feel are appropriate to them under the circumstances that they perceive are imposed upon them by their societal holding environment. That behaviour includes:

- All decision-making being done at the top with no discretion available to the lower levels. It is simply a matter of following the rules. In China if you were to seek to empower your lower-level employees you would be seen as a weak leader, one who provides no clear direction. Respect for authority is an important value which is reflected in the behaviour.

- A philosophy that stresses the importance of the family and of working like a team means that the emphasis regarding relationships between businesses is one which is based on personal relationships. In China it is important to build the relationship before conducting business. Thus, behaviour aimed at building trust is considered vital.

- The development of self-mastery toward a common goal means that the Chinese will listen to what the other party wants. Being considerate is important. In negotiating with other parties they will not be used to speaking up for what they want. Rather, they will rely on trust with the other side. In China, one's own personal authority and one's own personal trust are vital. In the absence of this behaviour by the other party, there is a strong chance that no business will be conducted.

This leaves us with the challenging task, which can be seen from either perspective, of how to develop a relationship between the individual and the outsider – the Confucian 6th relationship. To begin to answer this question, I shall start from the perspective of what happens when we are not able to share reality concepts with others. By exploring the notions of 'organizational socialization' and 'culture shock' we can gain valuable information about what happens to us when we operate in an alien culture.

## LEARNING FROM ORGANIZATIONAL SOCIALIZATION AND CULTURE SHOCK

The being that each of us infers from observing our own organism and behaviour and comparing it with that of others is what we call the 'self'. For example, I see myself as a consultant, a husband, a writer and a manager, and each of these aspects of my identity stems from my roles. I am also identified by my personal and bodily characteristics – punctual, even-tempered, dog-loving and balding. Finally, I have the attributes of my English nationality which may include modesty, personal privacy, independence, conservatism, and being non-emotional in public. Each of these characteristics defines both similarities with and differences from other people. They enable me to be identified and they indicate to the world and to me that I have a certain status, certain powers and responsibilities, and certain possessions that are essentially my own.

Clearly, my notion of 'self' relies upon other people with whom I have a relationship that enables me to be identified as 'me'. Part of the context in which I exist is through the organizational and institutional roles which I have adopted. Louis defines organizational socialization as 'the process by which an individual comes to appreciate the values, abilities, expected behaviours, and social knowledge essential for assuming an organizational role and participating as an organizational member' (1980, p. 229). Although developed over a longer period, we might define becoming a member of a society in much the same way. All the time we are in our own 'native' society and sharing the same basic assumptions with others we are at ease. However, when we enter an alien society the experience may be characterized by disorientation, foreigness and a kind of sensory overload.

Many writers refer to this experience as 'culture shock'. For example, Hunt refers to it as 'the experience of being a stranger in an alien and unfamiliar world' (1989, p. 33). She goes on to explain that immersion in an alien culture is an intense experience, and that most researchers report feeling some mixture of confusion, anxiety, excitement, frustration, depression, and embarrassment. Hunt also describes how some researchers have compared this experience to dying and that others report frequent anxieties about body health. Some of the discomfort stems from the fact that the researchers have lost their bearings, do not know how to communicate in their new setting, and often feel like helpless children.

It seems clear that the reason for 'culture shock' is, in its simplest terms, one of creating new boundaries. In order to gain a deeper understanding we need to explain what is involved in this process. As a convenient starting point we need to be aware that in every change there is a loss of the currently known. What happens to the personality when we go through a process of dramatic socialization? From the moment of birth a child is related to the world around him, and the roles that a person performs in life are made up of a complex series of focal action patterns which constitute *a repertoire of problem-solving solutions*. This repertoire, because it is based on experience, assumes that reasonable expectations of the world will be fulfilled. As time goes by, the individual's stock of 'solutions for all eventualities' grows greater, and novel solutions become rarer. Any major change will affect this situation. In such a case, the change not only alters expectations at the level of the local action patterns but also alters the overall plans and roles of which these form a part. Such a change is a huge upheaval and one which could result in disintegration – it is a world which is in chaos.

That I require the continuity, consistency and confirmation of my world that Kernberg (1966) refers to, is without doubt. If the possessions and roles by which we gain our continuity, consistency and confirmation are shared, then we can assume that if we lose our ability to predict and to act appropriately, our world will begin to crumble, and since my view of myself is inextricably mixed up with my view of the world, that too will begin to crumble. If I have relied upon other people or possessions to predict and act in many ways as an extension of myself, then the loss of those people or possessions can be expected to have the same effect upon my view of myself as if I had lost a part of myself. In other words, if I am in an alien holding environment I am without my usual support.

The missing external part of the holding environment is reasonably obvious to the individual in an alien culture but the internalized objects of the internal part of the holding environment still travel with us. These include, at the macro level, national values and attributions which were briefly referred to above and which will be significantly different depending on one's cultural background At the micro level, these will be more about those influences that derive from the culturally patterned behaviour of significant individuals, starting from when you were a child. Each of us has experienced different influences but by self-reflection you can ascertain what these are for you. Thus, the experience of working in a society where our stock of 'solutions for all eventualities' is found to be ineffective, presents us with huge problems. But provided we can be aware of our position we can do something about it.


## DEVELOPING CROSS-CULTURAL AWARENESS

From the previous chapters we have learnt of the many perspectives (both East and West) that we can take in trying to gain a deeper awareness of others. It has been suggested that by knowing the 'other' better we can achieve a more effective means of working. I feel quite certain that most of us would agree with this sort of statement in much the same way that we would agree with the virtues of 'apple pie', or, 'fried rice'. But, there is also a recognition of the complicated nature of the present-day multinational organizations. For example: 'We see frequent breakdowns in inter-alliance communications and we know that employees don't see eye-to-eye'; 'It is not a matter of proposing an homogenization of the world's cultures, rather it is a matter of learning to understand how and why people react as they do'; 'Much of our cross-cultural training and research occurs within the framework of bipolar cultural dimensions'; and, 'The management of human resources is becoming, for Joint-Venture enterprises (JVs), the battleground between performance and efficiency.'

Learning about the dynamics of life is a matter that has to be considered in the context of a relationship between two people, or two groups of people. As Gregory Bateson (1979), pointed out, it is correct (and a great improvement) to begin to think of the two people to the interaction as two eyes, *each giving a monocular view* of what goes on and *together giving a binocular view* in depth. This double view is the relationship. Relationship is not internal to the single person. Indeed, it would be nonsense to talk about 'dependency' or 'aggression' or 'prejudice' and so on as an individual activity. We cannot be dependent unless there is someone to be dependent on; we cannot be aggressive unless there is someone to be aggressive with; and, we cannot be prejudiced unless there is someone to be prejudiced about. All such words have their roots in what happens between persons, not in something-or-other inside a person.

Entering into a 'relationship' means that you do not have to lose your 'self' concept and nor does the other individual or group. Indeed, the nature of a relationship is a mutual recognition of sameness and difference that results in the binocular vision described above. However, before we can reach a position where we can see things through both eyes we need to escape from our own monocular vision. If we are so concerned and preoccupied with our own situation, because, for example, we are suffering from 'culture shock', we shall be unlikely to be able to get beyond thinking about ourselves. Where that is the case, we frequently use the other individual or group as a means of easing our own distress by, for example, stereotyping or scapegoating.

Stereotyping is essentially a form of defensive behaviour that we use to deny difference. Stereotypes may be defined as fixed, inflexible notions about an individual or group, and are at the heart of prejudice. They block our ability to think about individuals or groups as the 'other' (the outsider). Thus the usual approach is to concentrate on the 'differences' between 'us' and the 'other'. In doing so we accentuate the way we both behave, so that we draw boundaries and then hide behind those boundaries, taking some comfort that there are others not like us and that we are not like them. This applies to both 'us' and the 'other'. The difficulty is that in accentuating the differences we may also exaggerate them. Thus, for example, Eastern people may come to be seen, by Western people, as corrupt and the causes of all their problems; while Western people may come to be seen, by Eastern people, as immoral and the causes of all their problems. There is something satisfying, for both sides, in being able to say 'I am not like them'. By accentuating the difference we can disown parts of ourselves that we do not like.

In effect, this is what happens regarding scapegoating. Here, there is a displacement of, say, prejudice on to an individual or a group. It is easier to displace these impulses on to someone we consider 'corrupt' or 'immoral' than it is to displace them on to someone we regard as an equal. Therefore, when we are faced with personal or group problems which we cannot handle and which cause us to become increasingly anxious, one of the ways we deal with these feelings is to look to displace them on to someone else – a group or individual – and blame them for what is happening. Thus, it no longer becomes our 'failings' but their 'failings' which are the cause of the problem. This is something that all human beings do to a greater or lesser extent – if we are not aware of our actions.

This sort of self-satisfying activity prevents us from exploring the 'sameness' that clearly exists. To consider and accept the 'sameness' means that we have to give up our previous notions: for example, that others are 'corrupt' but not us, if we are of Western origin; or that others are 'immoral' but not us, if we are of Eastern origin. To accept the notion of sameness means that both sides have to give up deeply held views. When we accept the notion of 'sameness' it means that we no longer have these 'different' individuals or groups that we can blame for our inability. We now have to take our own authority for our actions – we have to 'own' not 'blame'.

The fact is that if we wish to develop the insights and attitudes which will further the development of both parties it is of prime importance to take an approach which considers what is involved in the situation for *each* party. This means that we need to understand that:

- My reality, or reality from my perspective, may be totally different from that of the other party. For example, as was described above, in Chinese culture to empower staff is a sign of weakness, whereas in the West it is accepted and encouraged practice.
- My 'obvious right answer' may not be at all 'obvious' or 'right' to the other party. For example, an assumption by someone from the West that they can make a deal with a low-level employee, as at home, will not work in China.
- Furthermore, it is not helpful to attempt to understand the other party independently from yourself. For example, in Chinese culture there is a need to build trust before a deal can be considered, and consequently, in business dealings it is necessary to understand how you, the other, are viewed.

If we make a serious effort at understanding ourselves in the context of a given situation, trying to see how we have contributed to it – willingly or unwillingly, consciously or unconsciously – then our view of the matter is nearly always altered, as is our manner of handling it.

A way of achieving this binocular view is to adopt a reflective approach, using the procedure detailed below:

- *Self-reflection.* By personal reflection on the way our own individual experience has been affected by the external and internal parts of our holding environments we will gain essential knowledge of 'self'. This will lead to:

- *Self-awareness*. Knowledge of self will provide an understanding of what, for us, is right or wrong, good or bad – the values, beliefs, customs, and ethics that guide our behaviour. This will lead to:
- *Self-control*. An ability to ensure that we do not use the other individual or group as a means of easing our own distress by, for example, stereotyping or scapegoating. This will lead to:
- *Awareness of others*. Once we have achieved sufficient self-awareness and an ability to control our disappointments and excitements we can then begin to learn about the 'other'. We shall now have a real opportunity for binocular vision.

To be able to move from a monocular to a binocular view requires self-awareness. This is unlikely to be gained without the active help of the other parties concerned. They need to *view* what is going on from both perspectives. This 'us' and 'other' double view is the relationship. Relationship is not internal to, say, the single Western person, or, indeed, the single Eastern person. In this context, it would be nonsense to talk about 'prejudice' from only one perspective. 'Prejudice' has its roots in what happens between persons, not in something-or-other inside a person. It is only when both sides *act together that they will have a binocular view* in depth. Most importantly, we shall be able to do the same for others. Without our help, 'others' will never be fully aware of the results of their behaviour. Without working across the boundary with 'other' people to develop a 'relationship' there can be no true appreciation of the effects of their behaviour. It is only by exploring 'shared behaviour' – not some abstract notion in a book, but real life behaviour and the feelings associated with it as it happens in a 'shared relationship' – that true understanding will develop.

For you this means taking your own authority for your own actions and working across the boundary in an assertive but non-threatening manner to provide feedback that explains your feelings resulting from the behaviour of the 'other'. You are the only ones who are aware of these feelings, and without making them conscious and public no one else will ever (truly) know what they are doing to you. For the 'other' person this means taking their own authority for their own actions and working across the boundary with a willingness to disclose their true beliefs and feelings with a view to developing their self-awareness and understanding.

This can only be achieved by a true relationship. 'Other' people are not going to give the necessary feedback to 'us' if they are not heard or accepted. But equally, we are not going to disclose data about ourselves if 'other' people are not able to accept our feedback. In the end it all depends on the willingness of both parties to be open and trusting with each other: for both parties to share their own monocular vision with a view to developing a binocular vision which permits a true relationship. In this sense we may say that it is not an 'us' *or* 'other' issue but an 'us' *and* 'other' issue – working together in a relationship. The key is to reflect on your own culture by working at the way it developed, so that you have an understanding such as in the example used above. You will then gain sufficient self-awareness to know what is you and what is not you. From this point, you will be able to become aware of others and their culture, and then be able to develop a more effective working relationship.

## References


Bateson, G. (1979) *Mind and Nature* (New York: Bantam).

Bhaskar, R. (1975) *A Realist Theory of Science* (Brighton: Harvester).

Hunt, J. C. (1989) *Psycho-Analytic Aspects of Fieldwork* (London: Sage).

Kernberg, O. (1966) 'Structural Derivatives of Object Relationships', *International Journal of Psycho-Analysis*, 47, pp. 236–53.

Louis, M. (1980) 'Surprise and Sense-Making: What Newcomers Experience in Entering Unfamiliar Organizational Settings', *Administrative Science Quarterly*, 25 (June), pp. 226–51.

Stapley, L. F. (1996) *The Personality of the Organization: A Psycho-Dynamic Explanation of Culture and Change* (London: Free Association Books).

Winnicott, D. W. (1971) *Playing and Reality* (Harmondsworth: Penguin).

# 10 How Westerners Can Make Sense of Asian Cultural Paradoxes

## Allan Bird and Joyce S. Osland

Oh, East is East, and West is West, and never the twain shall meet,
Till Earth and Sky stand presently at God's great Judgement Seat;
But there is neither East nor West, Border, nor Breed, nor Birth,
When two strong men stand face to face, tho' they come from the ends
of earth!

(Rudyard Kipling, *The Ballad of East and West*)

China's culture is rich with traditions and customs that reflect its well-established values of living in harmony with nature. Taoist notions of *yin* and *yang*, and architectural and landscaping practices predicated on *feng shui*, provide two such examples. Therefore, at a time when countries such as the USA, known for their orientation of 'mastery over nature', have moved away from building large hydroelectric dams, even dismantling some, why is China completing construction of a dam that will cover an area larger than some countries?

In Thailand, criticism is always considered – never constructive. Of course, there are differences of opinion, but *krengjai* – consideration of others – requires that people be quite cautious about openly expressing them for fear that expression will lead to conflict or be viewed as criticizing the views of another. Yet Thai politicians routinely denounce each other in public meetings; and their loyal lieutenants will speak for them in openly criticizing rival politicians.

According to Hofstede's (1980) research, the Japanese have a low tolerance for uncertainty while Americans have a high tolerance. Why, then, do the Japanese intentionally incorporate ambiguous clauses in their business contracts, which are unusually short, while Americans dot every 'i', cross every 't', and painstakingly spell out every possible contingency?

Virtually any Westerner with experience in Asia can identify similar examples, which contradict and confound attempts to neatly categorize Asian cultures. They violate conceptions of what Westerners think Asian cultures are like. Constrained, stereotypical thinking, however, is not the only problem. The more exposure and understanding one gains about any culture, the more cultural paradoxes, defined as situations that exhibit an apparently contradictory nature, one observes. For example, Japanese are widely believed to be polite and formal. Nonetheless, Westerners who ride a rush-hour subway in Tokyo will find themselves pushed and jostled, elbowed and crushed without a single 'Excuse me' to be heard. Long-term sojourners in Asia and serious cultural scholars struggle to make useful generalizations since so many exceptions and qualifications to the stereotypes, on both a cultural and individual level, come to mind.

Our frustration with the accepted conceptualizations of culture and how they are used led us to develop a model of cultural sensemaking (Osland and Bird, 2000). Our purpose in this chapter is to identify the Asian cultural profile, discuss models of progressive cultural learning, explain the cultural sensemaking model and apply it to cultural paradoxes in Asian culture. The basic sensemaking model works effectively in all cultures. It is particularly useful for Westerners working in Asia because of the frequency with which they appear to encounter baffling cultural paradoxes. Asian cultures are more culturally distant from Western cultures; this means that Westerners find them more difficult to decode and must work much harder at understanding and adjusting to them. Consequently, when Westerners are novices in Asia, they tend to apply simplistic, hard-and-fast rules to make sense of the Asian culture. Previously, we have referred to these rules as 'sophisticated stereotypes' (Osland and Bird, 2000).


## SOPHISTICATED STEREOTYPING

A look at the comparative management literature reveals that cultures are described in somewhat limited terms (Parsons and Shils, 1951; Kluckhohn and Strodtbeck, 1961; Hofstede, 1980; Ronen and Shenkar, 1985; Hall and Hall, 1990; Schwartz, 1992; Hampden-Turner and Trompenaars, 1994; Trompenaars and Hampden-Turner, 1997). There are 23 dimensions commonly used to compare cultures, typically presented in the form of bipolar continua as shown in Table 10.1. This table may reflect the Western propensity to think in terms of dualisms (Tripathi, 1988). An unanticipated consequence of using these dimensions, however, is the danger of stereotyping entire cultures.

In describing a disliked neighbouring ethnic group, the same kind of abusive terms – 'lazy', 'dirty', etc. – are used in most parts of the world. This is a low-level form of stereotyping, often based on lack of personal contact and an irrational

*Table* 10.1   Common cultural dimensions

| *Subjugation to nature* | *Harmony* | *Mastery of nature* |
| --- | --- | --- |
| Past | Present | Future |
| Being | Thinking | Doing |
| Evil human nature | Neutral or mixed | Good |
| Hierarchical | Collectivistic | Individualistic |
| Private space | Mixed | Public |
| Monochronic time | | Polychronic time |
| Low context language | | High context language |
| Low uncertainty avoidance | | High uncertainty avoidance |
| Low power distance | | High power distance |
| Short-term orientation | | Long-term orientation |
| Collectivism | | Individualism |
| Femininity | | Masculinity |
| Universalistic | | Particularistic |
| Neutral | | Affective |
| Diffuse | | Specific |
| Achievement | | Ascription |
| Individualism | | Communitarian |
| Inner-directed | | Outer-directed |
| Long-term | | Short-term |
| Hierarchy | | Egalitarianism |
| Embeddedness | | Autonomy |
| Mastery | | Harmony |

*Sources*:   Parsons and Shils (1951); Kluckhohn and Strodtbeck (1961); Hofstede (1980); Hall and Hall (1990); Schwartz (1992); Hampden-Turner and Trompenaars (1994); Trompenaars and Hampden-Turner (1997).

dislike of people who are different from oneself. We supplant one form of stereotyping for another when we view Koreans, for example, only in terms of Hofstede's cultural dimensions – moderate power distance, strong uncertainty avoidance, moderate masculinity and low individualism. In doing so, a complex culture is reduced to a shorthand description that people may be tempted to apply to all Koreans. We call this *sophisticated stereotyping*, because it is based on theoretical concepts and lacks the negative attributions often associated with its lower-level counterpart. Nevertheless, it is still limiting in the way it constrains individuals' perceptions of behaviour in another culture.

Do we recommend against using the cultural dimensions shown in Table 10.1 so as to avoid sophisticated stereotyping of Asians? Not at all. These dimensions are useful tools in explaining cultural behaviour. Indeed, cultural stereotypes can be helpful – *provided* we acknowledge their limitations. They are more beneficial, for example, in making comparisons *between* cultures than in understanding the wide variations of behaviour *within* a single culture. Adler (1997) encourages the use of 'helpful stereotypes', which have the following limitations: (1) they are consciously held; (2) descriptive rather than evaluative; (3) accurate in their description of a behavioural norm; (4) the first 'best guess' about a group prior to having direct information about the specific people involved; (5) modified based upon further observations and experience.

Westerners need to recognize that their original characterizations of Asian cultures are 'best guesses' that need to be modified based on further experience and contact. The limitations of sophisticated stereotyping become most evident when confronted with cultural paradoxes. This is the moment of realization that one's understanding is incomplete, misleading and potentially dangerous.


## THE ASIAN CULTURAL PROFILE

Let us begin with some 'best guesses' about Asian culture. For our purposes in this chapter, the term 'Asian cultures' refers to East Asia and Southeast Asia. While each Asian culture boasts a unique history, the region shares some aspects of historical development. For example, with few exceptions these cultures have been profoundly influenced by the diffusion of Buddhism throughout the region. Furthermore, the influence of China – its culture, traditional social institutions, and elements of Confucianism – extend across the region through longstanding economic and political relationships. Further, as a consequence of the Chinese diaspora, the overseas Chinese community is a potent force in every country in Asia.

When contrasted with one another, the individual cultures within this region are strikingly different. When viewed within the larger context of all world cultures, however, it is possible to view them as a group and refer to them generally as Asian cultures. Numerous multi-country studies of culture have found this to be the case (Hofstede, 1980; Ronen and Shenkar, 1985; Schwartz, 1992; Trompenaars and Hampden-Turner, 1997) although Japan often stands alone outside the Asian cluster. Table 10.2 demonstrates both the similarity and the differences across the cultures with regard to the 23 bipolar dimensions on which there is available data.

For purposes of making rough comparisons, we sifted through the sources cited in Table 10.1 and placed countries into one of three categories (at either pole or in the middle) based on their original scores. Though such an approach is neither methodologically rigorous nor highly accurate, the resultant picture is one that confirms the well-established perceptions about cultural similarity among Asian cultures. On most dimensions, a majority of national cultures cluster in a single category, e.g., high context language. When the several cultural dimensions from a single perspective were viewed collectively, the clustering appeared even more pronounced. For example, Schwartz found that three Asian cultures were highly similar on two dimensions, scoring high on embeddedness and hierarchy, and differ significantly only on the third dimension, mastery-harmony, as shown in Table 10.2.

*Table* 10.2   The location of selected Asian cultures on bipolar dimensions

| *Subjugation to nature* | *Harmony* | *Mastery of nature* |
| --- | --- | --- |
| | **C, J, T** | |
| Past | Present | Future |
| **C** | **I, Th** | **S, J** |
| Being | Thinking | Doing |
| | **C, HK, T** | |
| Evil human nature | Neutral or mixed | Good |
| | **T, HK** | **C, J** |
| Hierarchical | Collectivistic | Individualistic |
| | **C, HK, T** | |
| Private space | Mixed | Public |
| | **HK** | **C, P, Th, J** |
| Monochronic time | | Polychronic time |
| **C, HK, J, S, SK, T, V** | | **I, M, P, Th** |
| Low context language | | High context language |
| **S, SK** | | **C, HK, I, J, M, P, T, Th, V** |
| Low uncertainty avoidance | | High uncertainty avoidance |
| **M, S, V** | **C, I, P, T, Th** | **J, SK** |
| Low power distance | | High power distance |
| | **C, J, SK, T, Th, V** | **I, M, P, S** |
| Short-term orientation | | Long-term orientation |
| **I, P** | **S, Th,** | **C, HK, J, SK, T,** |
| Collectivism | | Individualism |
| **I, M, S, SK, Th** | **C, J, P, T, V** | |
| Femininity | | Masculinity |
| **Th** | **C, I, M, S, SK, T, V** | **J, P,** |
| Universalistic | | Particularistic |
| **M** | **I, J, P, S** | **C, SK** |
| Neutral | | Affective |
| **C, HK, I, J** | **S** | **M, P, Th** |
| Diffuse | | Specific |

*Table* 10.2    (*cont'd*)

| Subjugation to nature | Harmony | Mastery of nature |
|---|---|---|
| **C, I, J, SK** | **M, P, S** | **HK** |
| Achievement | | Ascription |
| | **C, HK, J, P, S, SK, Th** | **I** |
| Individualism | | Communitarian |
| | **M, TH, SK** | **C, I, J, P, J** |
| Inner-directed | | Outer-directed |
| | **SK, Th** | **C, HK, I, J, M, P** |
| Hierarchy | | Egalitarianism |
| **C, J, SK** | | |
| Embeddedness | | Autonomy |
| **C, J, SK** | | |
| Mastery | | Harmony |
| **C** | | **J, SK** |

C = China, HK = Hong Kong, I = Indonesia, J = Japan, M = Malaysia, P = Philippines, S = Singapore, SK = South Korea, T = Taiwan, Th = Thailand, V = Vietnam.

All of these studies are based on an etic perspective, which compares two or more cultures and yields a sophisticated stereotype. In contrast, studies with an emic perspective look at a culture from within its boundaries and identify its unique values. Therefore, emic studies identify greater differences between Asian cultures than do etic studies.

As managers work in different countries throughout Asia, they come to appreciate the diversity of Asian cultures and learn to distinguish among them. They also move beyond the first 'best guesses' to build a more layered, complex understanding of Asian cultures. There is a tendency for novices, however, to view Asian culture as monolithic, a tendency that is encouraged by sophisticated stereotypes. When novices confront cultural paradoxes, they need to make sense of the discrepancy between their observations and the sophisticated stereotypes they have learned.

## UNDERSTANDING ASIAN CULTURES

When Westerners look at Asian cultures, they inevitably interpret its institutions and customs using their own lenses and schemas; cultural myopia and lack of experience prevent them from seeing all the nuances of another culture.

In particular, a lack of experience with Asian cultures creates difficulties for Western expatriates trying to make sense of what they encounter. The situation is analogous to putting together a jigsaw puzzle. Though one may have the picture on the puzzle box as a guide, making sense of each individual piece and understanding where and how it fits is exceedingly difficult. As more pieces are put into place, however, it is easier to see the bigger picture and understand how individual pieces mesh. Similarly, as one acquires more and varied experiences in the new culture, one can develop an appreciation for how certain attitudes and behaviours fit the 'puzzle' and create an internal logic of the new culture.

The danger when Westerners employ sophisticated stereotyping in Asia is that it may lead individuals to think that the number of shapes that pieces may take is limited and that pieces fit together rather easily. As Barnlund (1975) notes, 'Rarely do the descriptions of a political structure or religious faith explain precisely when and why certain topics are avoided or why specific gestures carry such radically different meanings according to the context in which they appear.'

Westerners in Asia tend to focus first on cultural differences and make initial conclusions that are not always modified in light of subsequent evidence (Osland, 1995). Proactive learning about another culture often stops once a survival threshold is attained, perhaps due to an instinctive inclination to simplify a complex world. This is particularly likely with Asian cultures, which are characterized as high context – cultures in which contextual cues convey significant quantities of meaning (Hall and Hall, 1990). For example, smiling is an important element of Thai communication. Anthropologists have identified 18 distinctive Thai smiles, each with its own meaning – happy, discomfited, sad, embarrassed, and so forth. Regardless of the spoken word, it is the smile that conveys the real meaning of the communication. Westerners working in Thailand often find it difficult to distinguish among types of smiles and consequently misinterpret communication. Some never come to understand the importance of a smile; others find the variety of smiles overwhelming. This may lead them to seek black-and-white answers rather than tolerate the continued ambiguity that typifies a more complete understanding of Asian cultures and their high context.

One of the best descriptions of 'the peeling away of layers' that characterizes deeper cultural understanding is found in a fictionalized account of expatriate life written by an expatriate manager, Robert Collins (1987). He outlines ascending levels on a Westerner's perception scale of Japanese culture that alternate, in daisy-petal-plucking fashion, between seeing the Japanese as significantly different, and not really that different at all:

The initial Level on a Westerner's perception scale clearly indicates a 'difference' of great significance. The Japanese speak a language unlike any other human tongue . . . they write the language in symbols that reason alone cannot decipher. The airport customs officers all wear neckties, everyone is in a hurry, and there are long lines everywhere.

Level Two is represented by the sudden awareness that the Japanese are not different at all. Not at all. They ride in elevators, have a dynamic industrial/trade/financial system, own great chunks of the United States, and serve cornflakes in the Hotel Okura.

Level Three is the 'Hey, wait a minute' stage. The Japanese come to all the meetings, smile politely, nod in agreement with everything said, but do the opposite of what's expected. And they do it all together. They really are different.

But *are* they? Level Four understanding recognizes the strong group dynamics, common education and training, and the general sense of loyalty to the family – which in their case is Japan itself. That's not so unusual, things are just organized on a larger scale than any social unit in the West. Nothing is fundamentally different.

Level Five can blow one's mind, however. Bank presidents skipping through streets dressed as dragons at festival time; single ladies placing garlands of flowers around huge, and remarkably graphic, stone phallic symbols; Ministry of Finance officials rearranging their bedrooms so as to sleep in a 'lucky' direction; it is all somewhat odd. At least, by Western standards. There *is* something different in the air.

And so on. Some Westerners, the old Japan hands, have gotten as far as Levels 37 or 38. (Collins, 1987, pp. 14–15)

The point of Collins's description is that it takes time and experience to make sense of another culture. The various levels that he describes reflect differing levels of awareness as more and more pieces of the puzzle are put into place. Time and experience are essential because culture is embedded in the context. Without context it makes little sense to talk about culture. Yet, just as its lower-order counterpart does, sophisticated stereotyping tends to strip away or ignore context. Thus, cognitive schemas prevent sojourners and researchers from seeing and correctly interpreting paradoxical behaviour outside their own cultures.


## A MODEL OF CULTURAL SENSEMAKING

To make sense of cultural paradoxes and convey a holistic understanding of culture, we propose a model of cultural sensemaking. The model shown in Figure 10.1 helps explain how culture is embedded in context. Cultural sensemaking is a cycle of sequential events: indexing context, making

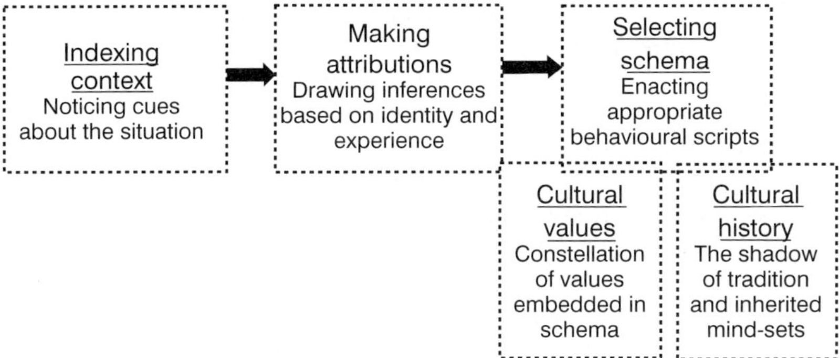


*Figure* 10.1    Cultural sense-making model

attributions, and selecting schema, which are under-girded by constellations of cultural values and cultural history.

- *Indexing Context.* The process begins when an individual identifies a context and then engages in indexing behaviour, which involves noticing or attending to stimuli that provide cues about the situation. For example, to index the context of a meeting with a Thai subordinate, we consider characteristics such as prior events (such as recent extensive layoffs due tothe regional economic crisis), the nature of the boss–subordinate relationships within and outside work in Thailand (relationships that often appear informal and easygoing to a Westerner, but that are rich with formality and hierarchy), the specific topic under discussion (raising employee morale), and the location of the interaction (boss's office).
- *Making Attributions.* The next step is attribution, a process in which contextual cues are analyzed in order to match the context with appropriate schema. The matching process is moderated or influenced by one's social identity (e.g., ethnic or religious background, gender, social class, organizational affiliation) and one's history (e.g., experiences and chronology). A senior Chinese manager who fought against the Japanese in the Second World War will make different attributions about context and employ different schema when he meets with a Japanese manager than will a South Korean manager of his generation, or a junior Malaysian manager whose personal experience with Japan is limited to automobiles, electronics, and sushi.
- *Selecting Schema.* Schemas are cultural scripts, 'a pattern of social interaction that is characteristic of a particular cultural group' (Triandis

*et al.*, 1984). They are accepted and appropriate ways of behaving, specifying certain patterns of interaction. From personal or vicarious experience, we learn how to select schema. By watching and working with bosses, for example, we develop scripts for how to act when we take on that role ourselves. We learn appropriate vocabulary and gestures, which then elicit a fairly predictable response from others.

- *The Influence of Cultural Values.* Schemas reflect an underlying hierarchy of cultural values. For example, Taiwanese subordinates working for US managers who have a relaxed and casual style and who openly share information and provide opportunities to make independent decisions, will learn specific scripts for managing in this fashion. The configuration of values embedded in this management style consists of informality, honesty, equality and individualism – values that are at variance with aspects of Taiwanese values. At some point, however, these same US managers may withhold information about a sensitive personnel situation because privacy, fairness and legal concerns would trump honesty and equality in this context. This trumping action explains why the constellation of values related to specific schema are hierarchical. It also helps us understand why cultural decoding by the Taiwanese subordinates may be so difficult.

- *The Influence of Cultural History.* When decoding schema, we may also find vestiges of cultural history and tradition. Mind-sets inherited from previous generations explain how history is remembered (Fisher, 1997). For example, perceptions about Indonesia's colonial era may still have an impact upon schemas, particularly those involving interactions with Westerners, even though a country gained its independence generations ago.

## SOME ILLUSTRATIONS OF SENSEMAKING

Sensemaking involves placing stimuli into a framework that enables people 'to comprehend, understand, explain, attribute, extrapolate, and predict' (Starbuck and Milliken, 1988). Let us analyze each of the cultural paradoxes presented in the introduction, using the sensemaking model.

Though it has a rich cultural history that emphasizes being in harmony with the forces of nature, with a population of over one billion people China struggles to address their needs and advance economically (indexing context). This situation is a source of some embarassment, as the Chinese see themselves as a superpower. Building such a large dam (schema selection) will help support the population, encourage further economic development and demonstrate to the world that China is superior to other nations. The dominant value underlying this schema is *face* (cultural value).

In the second example from Thailand, Thai politicians come almost exclusively from the upper class of a society populated by strong, established family-based factions. Political power is a mark of status and also provides opportunities to enrich oneself and the members of one's faction (indexing context). Denouncing opponents (schema selection) is part of a range of behaviours aimed at intimidating competing factions (out groups) who vie fiercely for a limited number of choice positions within the legislative and executive branches of government (schema selection). Loyalty to one's superior, and its counterpart, caring for one's subordinates (cultural values), override the value of *krengjai* (sophisticated stereotype).

In the third cultural example, when Japanese business people make contracts (indexing context), they (e.g., business people making attributions) opt for ambiguous contracts (selecting schema). The dominant value underlying this schema is collectivism (cultural value). In this context, collectivism is manifested as a belief that those entering into agreement are joined together and share something in common; thus, they should rely upon and trust one another. Collectivism trumps high uncertainty avoidance (sophisticated stereotype) in this context, but uncertainty avoidance is not completely absent. Some of the uncertainty surrounding the contract is dealt with 'upstream' in the process by carefully choosing and getting to know business partners and by using third parties. An additional consideration is that many Japanese like flexible contracts, because they have a greater recognition of the limits of contracts and the difficulties of foreseeing all contingencies (cultural history). Even though Americans are typically more tolerant of uncertainty (sophisticated stereotype), they value pragmatism and do not like to take unnecessary risks (cultural values). If a deal falls through, they rely upon the legal system for a resolution (cultural history).

**Taxonomy of Cultural Knowledge**

Sophisticated stereotypes provide a useful foundation for novices grappling with the initial stages of making sense of complex behaviours within cultures. However, rather than stereotyping cultures somewhere along a continuum, we can advance understanding by thinking in terms of specific contexts that feature particular cultural values which then govern behaviour. Geertz maintains that 'culture is best seen not as complexes of concrete behavior patterns – customs, usages, traditions, habit clusters – as has by and large been the case up to now, but as a set of control mechanisms – plans, recipes, rules, instructions (what computer engineers call "programs") – for the governing of behavior' (Geertz, 1973, p. 44).

In the evolution from novice to expert, people pay attention to different stimuli and develop different types of knowledge. For example, understanding the control mechanisms within a culture requires the acquisition of *attributional knowledge*, the awareness of contextually appropriate behaviour (Bird *et al.*, 1993). This is in contrast to factual knowledge and conceptual knowledge. *Factual knowledge* consists of descriptions of behaviours and attitudes. For example, it is a fact that Japanese use small groups extensively in the workplace. *Conceptual knowledge* consists of a culture's views and values about central concerns. Sophisticated stereotyping operates in the realm of conceptual knowledge. This category of knowledge is an organizing tool, but it is not sufficient for true cultural understanding. Knowing that the Japanese are a communal society (conceptual knowledge) does not explain the non-communal activities that exist in Japanese organizations or when the Japanese will or will not be communal. For example, why are quality control circles used in some work settings and not in others? Factual and conceptual knowledge about Japanese culture cannot answer that question; only attributional knowledge can.

Quite often, the knowledge about different cultures in Asia stays at the factual level and does not approximate or move on to the conceptual or attributional level. So we may know for example that the population of China is 4 billion and the population of Thailand is 60 million, but that provides no insight into cultural differences between the two countries. One reason that some Westerners do not progress far beyond the novice stage is that their cultural learning, like language learning, plateaus before complete understanding is achieved.

## Dialectical Model of Culture Learning

If we accept that cultures are paradoxical, then it follows that learning another culture often occurs in a dialectical fashion – thesis, antithesis, and synthesis. Thesis entails an hypothesis involving a sophisticated stereotype; antithesis is the identification of an apparently oppositional cultural paradox. Synthesis involves making sense of contradictory behaviour – understanding why certain values are more important in certain contexts. Behaviour appears less paradoxical once the foreigner learns to index contexts and match them with the appropriate schemas in the same way that members of the host culture do. Collins's (1987) description of the Westerner's Perception Scale in comprehending Japanese culture illustrates one form of dialectical culture learning, an upwardly spiralling cycle of cultural comprehension.

## WORKING FROM A SENSEMAKING APPROACH

The acquisition of cultural knowledge takes a good deal of time and energy, which is not available to all managers. So there are some trade-offs to developing attributional knowledge. Nor is it reasonable to expect employees who work with people from various cultures on a daily basis to master each culture. Nevertheless, organizing the knowledge they do acquire as context-specific schemas can speed up cultural learning and prevent confusion and errors in making sense of cultural paradoxes. The sensemaking model has clear implications for those who work across cultures, for organizations that send expatriates overseas, and for those who do research in Asia.

### Sensemaking for Individuals Working Across Cultures

After the training programme and once on assignment in a new culture, this cultural sensemaking approach has other practical implications.

- *Approach learning another culture like a scientist who holds conscious stereotypes and hypotheses in order to test them.* It is easy to get caught up in the exotic and mysterious, or to assume that Asian cultures are so different that one can only learn them through long years of exposure. One of the key differences between managers who were identified by their fellow MBA students as the 'most internationally effective' and the 'least internationally effective' is that the former group changed their stereotypes of other nationalities as they interacted with them, while the latter group did not (Ratiu, 1983). Active exploration and experimentation can both speed the learning process and give direction to what is learned.
- *Seek out cultural mentors and people who possess attributional knowledge about cultures.* Perhaps one of the basic lessons of cross-cultural interaction is that tolerance and effectiveness result from greater understanding of another culture. Making sense of a culture's internal logic and decoding cultural paradoxes is easiest with the aid of a willing and knowledgeable informant. This is particularly true for Westerners crossing into Asian cultures, which often contain fewer familiar cultural referents.
- *Analyze disconfirming evidence and instances that defy cultural stereotypes.* Even people with a great deal of experience in another culture can benefit from analyzing cultural paradoxes. For instance, the question of 'In what circumstances do Thais fail to exhibit smiles?' led to a more complex cultural understanding for one of the authors who had already spent much time in Thailand. Once expatriates can function reasonably well in another culture, it is easy for them to reach plateaus in their cultural

understanding and mistakenly assume that they comprehend the entire puzzle. This presents a danger when expatriates inadvertently pass on inaccurate information about the local culture or make faulty, and even expensive, business decisions based upon partial understandings. An Australian expatriate in Japan assured one of the authors that Japanese milk tasted different but was healthier because milk fats were replaced by fish oil. The reality was simply that milk in Japan had been pasteurized at a higher temperature (which altered the taste) to allow for a longer shelf life.

- *Learn cultural schemas that will help you be effective.* Knowing how to act appropriately in specific cross-cultural settings results in self-confidence and effectiveness. One cannot memorize all the rules in another culture, but understanding the values that underlie most schema can often prevent us from making serious mistakes.

## How Multinational Organizations Can Use the Sensemaking Model

The cultural sensemaking model also has practical implications for multi-national organizations.

- *Use cognitive complexity as a selection criterion for expatriates and people in international positions.* Particularly when it comes to selecting Westerners for Asian assignments, avoid black-and-white thinkers in favour of people who exhibit cognitive complexity, which involves the ability to handle ambiguity and multiple viewpoints. This skill is better suited to a thesis–antithesis approach to understanding the paradoxical nature of culture. As previously noted, the cultural distance between Asian and Western cultures is such that much greater flexibility is called for than when one is transferring managers between two Western cultures, say the UK to Finland.
- *Provide in-country cultural training for expatriates that goes beyond factual and conceptual knowledge.* Pre-departure cultural training is complemented by on-site training, which has the advantage of good timing. In-country culture training takes place when expatriates are highly motivated to find answers to real cultural dilemmas and when they are ready for greater complexity (Bird *et al.*, 1999). In-country training becomes all the more useful when the new culture contains greater complexity (high context) as is the case with most Asian cultures.
- *Act like learning organizations with regard to cultural knowledge.* Multinationals benefit from formal mechanisms to develop a more complex understanding of the cultures where they do business through such

methods as cultural mentors and in-country cultural training. There should also be mechanisms for sharing cultural knowledge. For example, getting returned expatriates to give formal debriefing sessions in which they report what they learned in their assignment increases the company's collective cultural knowledge and eases the expatriates' transition home by helping them make sense of a highly significant experience (Osland, 1995). Asian cultures are often seen as exotic or mysterious. In such situations it is likely that there is not only an absence of knowledge about a given culture, but also many misconceptions. One North American illustrated this point when, on hearing that a colleague was studying Thai in preparation for a new assignment, remarked, 'I've always wanted to visit Taiwan. Perhaps I could study Thai with you.'

## IN CLOSING

The words of Kipling at the front of this chapter were written at a time when Asian cultures were viewed as inscrutable – the Asian mind was unknowable. Yet Kipling held out hope that at some time in the future, both East and West would meet. The sixth generation is that time, and cultural sensemaking is one way that the twain can meet 'when two strong men stand face to face'.

## Acknowledgement

The authors would like to thank the UCLA CIBER Cross Cultural Collegium and ION for their contributions to the article. Dr Osland's research is partially funded by a grant from the Robert B. Pamplin Jr. Corporation.

## References

Adler, N. (1997) *International Dimensions of Organizational Behavior* (Cincinnati, OH: South-Western).

Barnlund, D. (1975) *Public and Private Self in Japan and the United States* (Yarmouth, ME: Intercultural Press) p. 6.

Bird, A., Heinbuch, S., Dunbar, R. and McNulty, M. (1993) 'A Conceptual Model of the Effects of Area Studies Training Programs and a Preliminary Investigation of the Model's Hypothesized Relationships', *International Journal of Intercultural Relations*, 17(4), pp. 415–36.

Bird, A., Osland, J. S., Mendenhall, M. and Schneider, S. (1999) 'Adapting and Adjusting to Other Cultures: What We Know But Don't Always Tell', *Journal of Management Inquiry*, 8, pp. 152–65.

Collins, R. J. (1987) *Max Danger: The Adventures of an Expat in Tokyo* (Rutland, VT: Charles E. Tuttle Co.).

Fisher, G. (1997) *Mindsets: The Role of Culture and Perception in International Relations* (Yarmouth, ME: Intercultural Press).

Geertz, C. (1973) *The interpretation of Cultures* (New York: HarperCollins Basic Books).

Hall, E. T. and Hall, M. R. (1990) *Understanding Cultural Differences* (Yarmouth, ME: Intercultural Press).

Hampden-Turner, C. and Trompenaars, A. (1994) *The Seven Cultures of Capitalism* (London: Piatkus Books).

Hofstede, G. (1980) *Culture's consequences: International Differences in Work Related Values* (Beverly Hills: Sage).

Hofstede, G. (1982) *Values Survey Module*, Institute for Research on Intercultural Cooperation, Netherlands.

Kluckhohn, F. and Strodtbeck, F. L. (1961) *Variations in Value Orientations* (Evanston, IL: Row, Peterson).

Osland, J. S. (1995) *The Adventure of Working Abroad: Hero Tales from the Global Frontier* (San Francisco: Jossey-Bass).

Osland, J. S. and Bird, A. (2000) 'Beyond Sophisticated Stereotyping: Cultural Sensemaking in Context', *Academy of Management Executive*, 14(1), pp. 65–87.

Parsons, T. and Shils, E. (1951) *Toward a General Theory of Action* (Cambridge, MA: Harvard University Press).

Ratiu, I. (1983) 'Thinking Internationally: A Comparison of How International Students Learn', *International Studies of Management and Organization*, 13, pp. 139–50.

Ronen, S. and Shenkar, O. (1985) 'Clustering Countries on Attitudinal Dimensions: A Review and Synthesis', *Academy of Management Review*, 10, pp. 435–54.

Schwartz, S. H. (1992) 'Universals in the Content and Structure of Values: Theoretical Advances and Empirical Tests in 20 countries', in M. Zanna (ed.), *Advances in Experimental Social Psychology* (New York, NY: Academic Press) 25, pp. 1–66.

Starbuck, W. H. and Milliken, F. J. (1988) 'Executives' Personal Filters: What They Notice and How They Make Sense', in D. Hambrick (ed.), *The Executive Effect: Concepts and Methods for Studying Top Managers* (Greenwich, CT: JAI Press) p. 51.

Triandis, H. C. (1982) 'Dimensions of Cultural Variations as Parameters of Organizational Theories', *International Studies of Management and Organization*, 12(4), pp. 139–69.

Triandis, H. C., Marin, G., Lisansky, J. and Betancourt, H. (1984) '*Simpatía* as a Cultural Script of Hispanics', *Journal of Personality and Social Psychology*, 47(6), pp. 1363–75.

Tripathi, R. C. (1988) 'Aligning Development to Values in India', in D. Sinha and H. S. R. Kao (eds), *Social Values and Development: Asian Perspectives* (New Delhi, India: Sage) pp. 315–33.

Trompenaars, F. and Hampden-Turner, C. M. (1997) *Riding the Waves of Culture* (London: Nicholas Brealey).

# 11 Cultural Literacy: Its Link to Business Success in Asia-Pacific

Philip Merry

## INTRODUCTION

I remember my first extended stay in another culture – that awful feeling of being an outsider, of seeming to be the only one who did not understand. It was in a small rural community in Sri Lanka. Most people had some English vocabulary, and so there was some potential for conversing, but the biggest difficulty was that there was a whole array of subtle cues and communication patterns, ways of connecting with the other person, that I was not at all part of and which disturbed me not just at the intellectual level but at a deeper emotional level. I felt in fact 'illiterate' and rather stupid and lost.

We have all been in situations like this where we felt illiterate, where there was a code of communication that everybody understood but us. We felt shut out, we felt somewhat at a loss not being able to see what was so obvious to everyone else. Language is obviously a factor which keeps us outside of a particular group, and that can be cured by learning the language; but it is the subtle signs of cultural illiteracy which are the most frustrating to deal with. Whilst this is uncomfortable at the social level, at the business level it can have a deep impact on efficiency and a corresponding effect on profitability. We may be at a business meeting or at a cocktail party, travelling with a potential partner, making a presentation, and we feel on the 'outside'. We hear words that we know, but we still cannot understand what is really being communicated between the others, or we cannot make ourselves understood.

I have lived in Asia for fourteen years and worked as a consultant and trainer in 33 countries helping organizations understand the impact of culture on their business. During this time I have developed a model for enhancing cultural literacy, but before examining the model let us take a look at some of the issues driving the need to understand culture in today's business world in Asia and indeed throughout the world.

# REASONS WHY CULTURAL LITERACY IS IMPORTANT

## Profitability: Soft Skills Mean Hard Cash

It is my experience that many organizations doing business in Asia do not pay attention to the soft issues – people and culture. There is still a tendency to put cultural issues second after the business issues – manufacturing, finance, sales, etc. It is fast becoming recognized that the so-called 'soft' skills are everything to do with profit – in terms of whether or not you get the deal, or whether you are operating with maximum efficiency. We often shy away from talking about money – but it is the impact of lack of cultural understanding on the bottom line that is driving a lot of the interest in cultural understanding. It is, of course, important to establish a supportive inclusive environment where all talents can be respected and valued, but nothing talks louder than the link to money and profit. Cultural understanding makes good business sense, and my phone rings more and more these days from people who recognize this, and who want make sure that their staff have the right skills. I always try to make this message loud and clear: if you are not taking account of cultural issues, you are losing money.

## Access to Information

What is not in dispute in today's world is that our understanding of events around the world – in cultures far away from our own – is greater than it has ever been at any time in world history. This is certainly true from the general world perspective. Internet and TV flash images to us and we can be in touch at the press of a button as never before. Want to connect with other people with similar interests? – just type in the defining word and suddenly you are connected to a global virtual community of like minds. To understand and communicate in the global marketplace one has to understand the values of those you are communicating with.

## The 'New' Economy

In the 'new economy', business has changed forever. Old historical markets are being changed, and the rules are totally new. The 'new' economy is making the managing of relationships more, not less, important. A key feature of the new economy is the outsourcing of many of the company's functions; and whereas control was the name of the game in the old economy because most processes are carried out by people within the company, now processes and products are not under the direct control of the company.

The managers and leaders of the future need to be good at influencing and relating to people who are in charge of major parts of their company's processes, who do not work directly for the company, and who are more often than not from different cultures.

## Massive Increase in Partnerships and JVs

Companies realize that the way forward in the new economy is to build up networks and partnerships and develop the synergy that helps to promote both parties' interests. You cannot open any newspaper without reading about a new merger, JV (joint venture), or take-over, but the defining feature is the need to work efficiently and harmoniously with people who often see the world in a different way from you. This has been amply expounded in the excellent recent study of JVs and partnerships (Spekman *et al.*, 2000).

## Corporate Culture vs. Local Culture

Many companies' values and policies reflect the values of the head office national culture, and these are often in conflict with the national cultural values of the various countries in which the company operates. Much of my work comes from the need for head office to adapt their corporate policies to local values.

## Teams Made Up of Many Cultures

It is a long time since I can remember working with a team which was mono-cultural. There are an increasing number of people who are working for a majority of their time in cultures other than their own. When you are constantly operating with other cultures – selling, designing, marketing, putting together financial packages – if you are not culturally literate the consequences soon make themselves known.

## Ease and Speed of Electronic Communication

What did we do before electronic communication? There are none of us who do not feel the benefit of being able to communicate so quickly. It does not seem so long since we were marvelling at the fact that it took only 10 seconds to fax across the world, and now we communicate instantaneously! But what I have noticed is that there are an increasing number of cross-cultural conflicts as a result of email. Issues that previously would have been dealt with sensitively in face-to-face situations are being dealt with a lot more directly electronically because it is easier to zap a critical, angry reply. This has highlighted the need for more effective interpersonal and cross-cultural skills, and the need to distinguish between when and when not to communicate by email. In a recent session for the senior management team running a factory, the head of manufacturing and head of marketing had a serious personality and cultural clash. As I listened to the argument about the exchange of angry emails I asked them how close their buildings were to each other. I was shocked to discover that their offices were next door to each other! When there is anger and upset it is always better to deal with the cultural issues face to face.

**Western Management and Leadership Development Theories**

If you look at the management section of any bookshop in Asia, 95 per cent of the texts you see will come from the West, and most of them from the US. Whilst there are many great management theorists writing in the West, let us not be blind to the fact that most of them are writing from a value perspective that cannot be applied indiscriminately in other cultures. A way to assess any new text is to look for the author's willingness to acknowledge their own cultural roots and values in their writings. The way people are motivated, view leadership, solve problems, handle conflict, brainstorm, lead teams, or strategize looks different according to the cultural lens through which you look. If these are some of the issues indicating the importance of cultural literacy, what is the model to help develop the skills?

## THE 4-STEP CULTURAL LITERACY MODEL

Essentially cultural literacy is the ability to read and respond to the signs of the other in an appropriate way. Too often discussion about culture gets left out of management discussions because it is seen as a vague and nebulous area which is all about personal opinions and preferences rather than objective fact. There can be some truth in this view because management models can never explain away the mystery of being human, and we are often too quick to put people into boxes. Yet we still need to have a language to talk about this complex issue, and over many years of consulting and training in the multi-cultural arena I have developed a relatively simple model to aid understanding when working with this issue – maybe it's because I have worked with a lot of engineers and they love models. My model draws heavily on the wisdom, advice and input of others, and highlights the need for four skills to develop cultural literacy:

- *Cultural 'Detective'* – seeking out and identifying cultural issues, and asking What is culture? What are the three levels of culture? What are the stages of development in multi-cultural relationships?
- *Cultural 'Auditor'* – analyzing cultural dimensions using the 7D model of Trompenaars/Hampden-Turner.
- *Cultural 'Integrator'* – integrating, reconciling opposite cultural issues.
- *Cultural 'Connector'* – improving 'face-to-face' cultural skills.

## DETECTIVE SKILLS

It is my experience in business that culture is the last thing to receive attention; it is often pushed under the carpet until it is nearly too late. We need to heighten our ability to detect the problems before they cause real damage to our teams and organizations. There are three things we need to understand in order to develop detective skills: What is culture?, What are the levels of culture?, What are the stages of development in multi-cultural relationships?

### What is Culture?

Culture has sometimes been described as 'the way we do things around here'. It does however hold different meanings for different people – some regard a 'cultured' person as one who enjoys art and opera. But the definition of culture goes much deeper. Culture comes from the Greek word 'Cultura' – meaning, how we interact with nature. Humans developed in different geographical situations and faced different problems. The responses that were produced to solve environmental problems were the first basis of cultural differences. Over time, these responses became so automatic that they dropped from our awareness and became core assumptions about life. The definition of culture we will use is: 'Culture is a group's shared system for solving problems'. Although our focus is on national cultures it is important to remember that there are many types of culture – professional, national, educational, ethnic, and corporate, plus many more.

### Three Levels of Culture

In order to correctly detect what the cultural problems are, we need to be clear about the different levels of culture. Most people jump to judgment when confronted with cultural issues, and therein lies the major reason why bad feeling develops between cultural groups. Recognizing the different levels of culture helps us detect the real problems more accurately.

## (a)    Level One – Observable Differences

What can help in detecting cultural issues is to first describe the observable differences, for by doing this it helps to remove the judgment that leads to problems. For example, when one asks an English person to describe which side of the street they drive on in America, they will often reply 'the wrong side'. They do not say 'they drive on the opposite side to us, I wonder why that is'. They will jump to a judgment that because it is different from the English way it is therefore wrong. When detecting cultural problems it helps to recognize whether people have moved to judgment or are just in the descriptive mode. Some observable differences are:

- food
- skin colour
- gestures
- expressions of emotion
- eating habits
- customer service
- clothes
- language
- sayings
- treatment of strangers
- driving patterns
- climate

For example, what is one of the observable differences when working with the Japanese? I have asked this question to hundreds of people and always the answer is that 'they bow'. There are other cultures that bow but none that bow so much as the Japanese – this is an observable difference. If you then ask a Japanese person 'why do you bow?' she or he is likely to answer that it is a sign of respect, i.e., it is a value or a norm. This brings us to level two of culture – values or norms.

## (b)    Level Two – Norms and Values

Observable differences are indications of values or norms; it is important to distinguish between the two. Norms are things that you do not necessarily believe, but you do them because everybody else does; values are things you really believe in. When trying to motivate people it is important to focus on their values, not on the norms they follow because everyone else is following them. The following case illustrates this point.

*Case Study: 'It's All About ME'*

I was asked to work with an American MNC in the food industry – let us call them 'Maxdor Enterprises'. They were implementing an employee improvement programme in the US and wanted to adapt the scheme to Indonesia and Malaysia. The scheme was going well in the US and was called 'It's All About ME'. This was a clever and motivating title because it linked directly to people feeling they could make a difference, and it was also an acronym for the name of the organization.

My first input to the planning team was to say that if the scheme was to be successful in Malaysia or Indonesia then it was important to look again at the title and maybe change it. Why would I say that and what has it to do with values and norms? Well, first of all, the scheme was going well in the US because the title of the scheme was motivating – it linked to a good American cultural value, 'individualism'. The difficulty is that when detecting cultural problems your own assumptions blind you to the fact that there may actually be cultural problems. Because most Americans are motivated by individualist values it is often forgotten that not all cultures share the same values. Most Asian cultures are more focused on group harmony, and in fact the scheme 'It's All About ME' would become known as the 'Selfish Scheme' in Malaysia. What is seen as a value in the US is only a norm in Asia, with significant negative consequences for important projects. The Malaysian HR manager told me that the scheme was expected to save US$10 million per year, but that if the original title was kept 50 per cent of her staff would not volunteer to join in. There is a potential loss of $5 million because of four words.

### (c)    Level Three – 'Hard Wired' Core Assumptions and Beliefs

Let us return to our Japanese colleague mentioned earlier, and say to her 'We see you that you bow and you tell us that it indicates respect in your culture – but why is respect so important in Japan?' She will probably be unable to answer, and if she does, she will probably say, 'It's our culture'. All of us have basic beliefs or core assumptions that we grew up with, and that our religion, education, parents, geography, heroes, or whatever, gave to us. These core values are automatic, things we take for granted, and they are so imbedded in the way we look at the world that we are often not even aware of them until we are in a situation where we meet others who see the world differently. In computer terminology these basic assumptions are our 'operating systems'. When 'detecting' cultural problems it is vital that we understand our own and others' hard-wired assumptions and beliefs. Most Western executives working in Asia have very different basic assumptions from their Asian colleagues, which if left unexamined can cost the business dearly.

## Stages of Development in Multi-Cultural Relationships

Finally, when detecting cultural problems we need to have a clear understanding of the stages of development of multi-cultural relationships. The four stages are discussed below and summarized in Table 11.1.

- *Stage 1: Starting/Hiding/Concealing.* Cultural differences are experienced at this stage, and either or both sides may be upset with their first impressions of the behaviour and perceived attitude of the other. These opinions are, however, kept hidden from the other culture, with both sides complaining about the illogical, disrespectful, rude, behaviour of the other side in private to their peers. Open sharing of difficulties with each other is avoided.
- *Stage 2: Opening/Sharing/Confronting.* A crisis stimulates open discussion of the hidden cultural issues. Both sides insist that their view is the correct one, and a stalemate can often result, with people blaming each other for the problems.
- *Stage 3: Improving.* At this stage there is a healthy recognition that there are different views and values, and a willingness by both sides to work together. Both sides emerge from the infighting and try to establish rules, guidelines and improvement plans in order to work more effectively together. There is an acceptance of the different perceptions that each have about the issues, conflicts are worked out in a positive way, and there is more cooperation. An outside facilitator or an enlightened manager/ leader is often needed to help with this stage.
- *Stage 4: Excelling.* Respect for the views of the other is high and there is a willingness to understand the other's point of view, and develop inclusive solutions. At this stage the team feels comfortable with each other; there is a tolerance for each other's strengths and weaknesses and an accepted way of doing things which has been tried and tested. The strengths of the different cultures are being used to the full and there is a feeling that they can really begin to achieve results.

## Recognising the Level

In order to detect cultural problems it is important that there is a level of sharing of those problems, but it is often difficult for people to share issues, for a variety of reasons. What is crucial when detecting the issues is to recognize

*Table* 11.1   Stages of team development: overview

| Starting | Sharing | Improving | Excelling |
| --- | --- | --- | --- |
| ♦ Cultural differences are experienced at this stage.<br>♦ Both sides may be upset with the behaviour and perceived attitude of the other.<br>♦ Opinions are however kept hidden from the other culture.<br>♦ Open sharing of difficulties with each other is avoided.<br>♦ *Output low.* | ♦ Crisis stimulates open discussion of the hidden cultural issues.<br>♦ Both sides insist their view is correct.<br>♦ Stalemate can often result with people blaming the other.<br>♦ Output is still low. | ♦ Healthy recognition that there are different views and values.<br>♦ Willingness by both sides to work together.<br>♦ Beginning to work to find integrating solutions.<br>♦ Output increases. | ♦ Respect for the views of the other is high.<br>♦ Willingness to understand<br>♦ Development the other's point of view. of inclusive solutions.<br>♦ Work output high.<br>♦ Recognition that solutions need to involve both parties.<br>♦ Output high. |

that people often push issues under the carpet, and they only come out when there is a crisis – which may be very damaging in many ways. So what is important is to help detect and deal with the problems before they do become too damaging. Finns have a good phrase for helping understand the stages. They talk about 'putting the cat on the table'. Leaders and facilitators need to be able to learn how to help 'put the cat on the table'. I am often asked to work with groups who are stuck at level one – two camps have formed and they have underlying resentment towards each other which is not being expressed. Helping the groups to find ways of addressing the issues is vital.

**Benefits of the Model**

1. It helps members see that there is a pattern to development and that confusion and conflict are a normal part of building effective multi-cultural relationships.
2. It gives a model against which the progress can be measured.
3. By understanding the cycle the parties can take an active approach to managing each stage of their development.
4. It legitimizes talking about cultural issues.
5. It helps detect problems before they become too damaging.

Once we have 'detected' the problems we need to find a language to describe them. We need an 'audit' framework.


## AUDITOR SKILLS

Whenever there are things wrong with a project we do a technical audit; if there are budget problems we do a financial audit. What surprises me is that we know that a lot of the problems come from the cultural or people area, yet in many years of consulting I have rarely found any team or organization which does regular cultural audits, to identify issues and find a way of solving them.

To audit is to find out which cultural dimensions are having an impact on the situation. People often avoid talking about cultural issues because they do not have a language which is clear or free from judgments. We need a way of looking at the issues once they have been detected, and obviously also need a language that has popular appeal, a solid basis in research, and intellectual credibility. For me, the work of Dr Fons Trompenaars and Dr Charles Hampden-Turner has all three (see, for instance, Trompenaars, 1993; Hampden-Turner and Trompenaars, 1997). I have for many years worked with the Trompenaars/Hampden-Turner 7D model, for the following reasons:

1.  The 7 D model is inclusive – most cultural issues can be audited using the 7 Dimensions.
2.  It is a model based on solid research.
3.  The data base is continuously updated, which reflects the changing nature of culture.
4.  Most of all, in the many workshops I have run using the model it seems to have more appeal to the real problems of real managers than other models on the market at a practical and useable level.

The model is described extensively in *Riding the Waves of Culture* (Trompenaars, 1993); what follows is a brief summary of the 7 dimensions.

### The Trompenaars/Hampden-Turner 7 Dimensions of Culture

1.  *Rules*: One universal rule or adaptation to particular circumstances?

*Universal* cultures tend to think there are rules which apply in every situation – the 'morally' right imperative. In business this shows itself with head-office rules which must be adhered to all over the world: 'We designed the best product or performance appraisal system and we have to have universal conformity.' *Particular* cultures see the obligation first to the particular set of circumstances or special relationships they face. Rules need to be adjusted to suit these special circumstances.

2.  *Work behaviour*: Individualistic or collective harmony?

*Individualist* societies uphold the rights of the individual, important issues being a concern for freedom, individual rights, and what individuals personally think and want. The mature person 'stands on their own two feet' and is independent. *Group*-oriented societies subvert individual need in favour of the good of the group as a whole. Problems often occur in teams and performance systems when these cultures clash. In meetings, individualists generally adhere to the philosophy of who shouts loudest holds the sway; Asians often defer to others in the group and do their utmost to maintain group harmony. Individualists focus on individual responsibility when things go wrong; whilst group-oriented people look to the group as a whole.

3.  *Emotions*: Freely expressed (affective) or more reserved (neutral)?

We all have emotions – the issue is how we express them. Some cultures are *neutral* and will keep their feelings to themselves. Other cultures are *affective* and think of it as a virtue to tell you exactly how they feel, and are suspicious of people who will not 'call a spade a spade'. In business dealings this can lead to enormous frustrations, with accusations of not telling the truth passed backwards and forwards. Learning how your partner expresses their feelings is extremely important when you spend a great deal of time in face-to-face contact with them.

4.  *Level of involvement*: Partial (specific) or whole (diffuse)?

*Specific* cultures tend to get involved with others on a partial specific basis and will give their opinion on a limited specific subject, e.g. this is my view as a manager, as an engineer, as a mother, etc.; life is a series of specific roles or boxes which are not connected to each other. This person can be involved in an argument with a subordinate and then be friendly with them in a private capacity because the argument was only in their work role, which does not impact on their role as friend. *Diffuse* cultures on the other hand have a more holistic view of life: there is no disconnect between the various areas of life; deep relationships where one shares oneself are valued.

5.   *Status*: What you are (achievement) or what you do (ascription)?

In *Achievement* cultures you are only as good as your last quarterly results – you are what you have achieved. In *Ascription* cultures status is given according to your position in society, age, sex, race, job title, etc. Western companies appointing to senior positions often appoint the 'best brains', most 'go-getting', someone who has 'achieved' in their career so far. Asian societies give far more respect to age than do Western societies, and hence, if a company appoints an able younger executive to manage older people there can be trouble if not managed properly.

6.   *Time*: Sequential or sequential?

People structure time very differently. For some people, 9.00 a.m. means 9.00 a.m., for others it could mean 10 (some call it rubber time). *Synchronic* people can do many things at the same time and are not so strict with the sequences of time. Their desk may look chaotic but there is an order that they recognize. *Sequential* people generally only do one thing at a time and keep order in their life – their desk is often clear and tidy. The trouble comes when these two styles work together. The key to managing their relationship is to recognize that both have their place and to use the strengths of each style.

7.   *Control*: Internal or external?

Most Western societies are *internally* focused and live as if they control their own destinies, and believe that results are based on their own effort. Most Asian cultures believe much more in destiny or fate having importance, i.e., things outside of their control. In the West, older patterns of religious belief have broken down and direction comes from within: 'I control my own destiny.' Asian societies give as much importance to *external* as to internal issues, and will say that religious beliefs and traditions play an equal part in controlling destiny. It is not a question of who is right or wrong, but one of understanding the other culture's differing perceptions.

Once we have 'audited' the patterns we need to be able to 'integrate'.


## INTEGRATOR SKILLS

It is vital to be able to move past audit to integration. If you stay at audit there is just a description of the problems with no resolution. Integration requires finding solutions where both points of view can be included – what Trompenaars and Hampden-Turner call a cultural reconciliation.

Integration or reconciliation requires knowledge and understanding of one's self and one's own values, knowledge and understanding of the values of the other, suspending of judgment and demonstration of respect in mediation of an integrating, reconciling and inclusive solution.

'Detecting', 'auditing', 'integrating' skills need to complemented by face-to-face 'connecting' skills.


## CONNECTING SKILLS

One can be expert in the first three skills but still not be able to solve cultural problems. This is because detecting, auditing and integrating skills are conceptual skills which can be understood at an intellectual level without needing to communicate with the other. Connecting is about face-to-face skills when the different parties meet together – skills which establish trust. You can have the best strategies, the best products, and the best technologies and R&D possibilities, but if the people who need to work together do not trust each other then all of those great competitive advantages are worthless. Building trust is to do with the ability to get inside the skin of what is important to the other, and to deliver results that make them *feel* good. To do that you have to understand their 'culture' – or their way of doing things. You have to be literate enough to read their world and the signs that cause them to feel good. To be effective in connecting skills you need to:

- understand your own emotions
- pay attention to first impressions
- understand your own personality
- establish rapport with the other
- know how to build trust
- understand body language and gestures
- *be respectful – transmit positive regard, encouragement and sincere interest*
- be non-judgmental; avoid moralistic value laden statements
- be aware of influence of your own values, perceptions, opinions, on human interaction; see them as relative rather than absolute
- display empathy; put yourself in the shoes of the other, seeing the world through the eyes of the other
- talk and listen; take turns in talking; observe speech patterns of other
- tolerate ambiguity

- understand conflict-handling patterns across cultures (The best study of the different conflict-handling patterns in Asia can be found in Leung and Tjosvold, 1998. This excellent book has illuminating chapters by different authors on conflict management in the Asian cultures.)

We have outlined the four skills of cultural literacy but how do we apply them in real business situations in Asia-Pacific? What follows are real cases of cultural situations that I have worked with in Asia with some of the world's most successful companies. Names and industries have been disguised. What we will do is to use the four cultural literacy skills to explain and resolve the cultural issues.

## WORKING WITH THE 4 SKILLS OF CULTURAL LITERACY

I will describe the cases and then apply the 4 skills to the cases.

- *Detecting* – Do we have a cultural problem?
- *Auditing* – Which of the 7 Dimensions are having an impact on the situation?
- *Integrating* – How to bring the alternative views together?
- *Connecting* – What are the ways to make sure that there is a good face-to-face rapport?

### Case 1: Dutch–Singapore Communication

'Keystroke' is a Dutch company producing personal hand-held computers. It is interested in a joint venture with a Singapore company, 'Orchid', to produce a new PC initially to be launched in Singapore and then Asia-wide. The Dutch engineers were invited to Singapore for an initial meeting to discuss the new project. The Dutch were the experts in technology and the Singaporeans were experts in design. This is a vital project as most PC manufacturers are preparing new models in this category.

After the first meeting, which was very friendly, the Singaporeans agreed to work further on the design. A month later they submitted their design to the Dutch in advance of the second meeting. The Dutch reviewed the proposal and thought that it was a new and brilliant design. However, there were several detail mistakes concerning size – the innovative new casing design was too small for the internal technical requirements. The Dutch wondered how to tell the Singaporeans.

The second meeting was called to finalize the design. The meeting was initially very effective, but after a while there were difficulties. Problems first began when Gert, a Dutch engineer, criticized the Singapore design, saying that the design 'lacked basic engineering common sense'. The Singaporeans were polite but seemed rather tense and answered the Dutch questions in a non-committal way. The Dutch began to be irritated by the silence saying directly that the Singaporeans were being 'inflexible'. The Singaporeans continued to seem uncooperative in the Dutch eyes as they would not join in direct discussion about how to improve the design. The Dutch had omitted to tell the Singaporeans that they liked the basic design.

Both sides left the meeting without agreement and feeling frustrated. Collaboration began to slow down, schedules were missed and the whole project was at risk.

The managing directors of both companies have heard that there are problems due to a 'cross-cultural misunderstanding'. This is an important project and they have ordered the two sides to sort this out.

*Applying the 4 Skills of Cultural Literacy*

(1) *Detect.* What are the cultural problems? This is a cultural problem as both sides are upset with the communication style of the other. When there is a silence from one side and an insistence from the other this generally indicates a cultural problem. Some people say that Gert's comment that the design 'lacks basic engineering common sense' would be rude anywhere. This is probably true, but the difference is that if Gert was speaking to a fellow Dutch, the other person would reply directly complaining about Gert's tone of voice and rudeness, and they would sort it out there and then.

(2) *Audit.* Which cultural dimensions are influencing this situation. The major dimensions here are:

- Individual–Collective
- Affective–Neutral

The Dutch are much more individualistic and affective than the Singaporeans. If they have a problem they will generally be straightforward in telling what the problem is. This would take place in the group setting where speaking out your own opinion is OK. The Singaporeans, on the other hand, are more collective in their style, and maintaining face and group harmony is encouraged. (But this is beginning to change as Singapore is more and more influenced be Western MNCs. Many Singaporeans have told me that they behave in a more Western way at work and Asian at home.) If there are critical things to say they will be said in a diplomatic way, and certainly there will be some praise as well. As far as feelings are concerned the Dutch began to be irritated be the silence of the Singaporeans and said so. The Singaporeans were also feeling upset, but being a more neutral culture, kept their feeling to themselves.

(3) *Integrate.*  What is important in integrating this situation is that both sides move towards each other's style, the Dutch needing to be more diplomatic, and the Singaporeans more direct. But before that can happen there needs to be an understanding and respect from both sides for the style of the other. It is very easy in these situations where there are basic communication problems for judgment to creep in. The Singaporeans would call the Dutch style rude, and the Dutch call the Singaporeans uncooperative. So real understanding and particularly the suspension of judgment is crucial for integration to take place.

(4) *Connect.*  What face-to-face approach needs to be taken in the meeting? What is of real importance here is that signs of respect are shown. I have discussed this case many times in every culture in Asia, and there is generally a view that what is needed more than anything is an apology from Gert. This allows collective cultures to recover 'face'. Many Europeans argue during the discussion that an apology is not necessary because Gert is just pointing out the truth and he has done nothing wrong. Gert needs to develop an ability to sound genuinely sorry for his words at the previous meeting; he has to pay attention to his tone of voice. What also often happens is that both sides can get caught up with the technical details and avoid the cultural issues. As far as connection skills are concerned from the Singapore side, they too need to show a conciliatory tone and demeanour if the project has a chance of success; this is shown by being willing to talk about the issues.

*Business Impact*

The consequences of not sorting this out are that it will cause a slipping of deadlines and efficient working together – and the cost of that could be enormous if it means that they do not get to market first with their new design. (For further understanding of Singapore's unique business culture, refer to Shein, 1996.)

**Case 2: Promotion in Hong Kong**

*'Highlands Bank'* is a British bank with operations in Hong Kong (HK). David Drew, from Yorkshire, England, has been manager of the corporate technology department for two years now. He has done well – the technical skills of his staff have developed greatly since he arrived, and he is generally well liked by both expatriates and locals. One issue that continues to be

raised is the promotion of locals. This is increasingly important with the return of Hong Kong to China – it is company policy to appoint HK Chinese to senior positions. David maintains his view that there are not sufficient staff with leadership potential in the HK office to take over from him, and he sees the need for the foreseeable future to have expatriates in the post. When questioned as to how he comes to that decision, he says that there are two who have potential – Peter Tan and Wendy Wong – but they do not have the necessary 'leadership' skills. As a way of testing their ability to speak up, David says that at his weekly meetings he constantly asks for new ideas or criticisms, but is met by silence – no one speaks up. He has come to the conclusion that the Chinese prefer others to take the lead as they are quieter by nature, and that the expats are better at speaking out. One of the major criteria of the job is to represent the region at meetings in London, where it is important to speak up or the others (mostly European) will dominate. He does not think his two supervisors would speak up.

*Applying the 4 Skills of Cultural Literacy*

(1) *Detect.*   What are the cultural problems? When there is behaviour that we find puzzling it is generally an indication of a cultural problem. It cannot be that there is no leadership among the Chinese as a nation, otherwise they would not occupy the position they do in the world. David's position is clearly ridiculous, but there are many Western managers who would no doubt jump to the same judgment. Western managers and trainers get caught up in this type of problem: they ask a group for their opinion and are met by silence, and assume that the people with whom they are speaking either do not have an opinion or, if they do, lack the courage or initiative to speak out.

(2) *Audit.*   Which of the 7 cultural dimensions are influencing this situation? The major dimensions impacting this situation are:

- Individual–Collective
- Achievement–Ascription

Many of the dimensions are linked, and are similar in their essence. In general, when you have individualist cultures it is often the case that they are also achievement-focused too; likewise most collective cultures are also ascription-oriented. The Chinese would not offer criticism to their boss in public, and that is what David is really asking his staff to do in the meeting. The basic issue is that David is transporting a dynamic which would work in Yorkshire, England, to Hong Kong. In Leeds, if David were to ask for some criticism at a departmental meeting, people would offer their opinions – some less directly that others, but he would get input. If he were looking for a successor where an important leadership criterion was to be outspoken, he would use the meetings to form an opinion of the people who could take over from him. All other skills being equal, the ability to be blunt and outspoken, 'to call a spade a spade', would be a major deciding factor in his choice of a successor. Individualist and Achievement oriented societies do not mind giving their opinion in public. In Hong Kong the Chinese are more Collective and Ascription focused – the boss is the boss and you accord her or him respect that their age and position demand, especially in public situations. So when there is no reply to David's questions it is not because the Chinese have nothing to say or that they lack initiative, but because of the context in which they are being asked to speak out. David's 'universalist' assumptions – i.e., if it is true in Yorkshire it is true in Hong Kong – lead him to a conclusion which would be very damaging to the morale of the local staff, and in the long term very costly in financial terms to the company.

(3) *Integrate.*   How to prepare a strategy for dealing with the succession planning issues? If the context is causing the Chinese problems then the integration is to observe the two supervisors in a different context. Wendy and Peter are already supervisors so they must already have some leadership qualities. David needs to observe them in their own team meetings and see how tough and outspoken they can be. It is important however that David does not attend the meetings himself, otherwise he will get the same reverential behaviour he gets in his departmental meetings. In this real situation that I dealt with we found that Peter was very mild and unassertive and did not have the ability to be tough when necessary. Wendy, on the other hand, ran her meetings with an iron rod, often raising her voice to deal with those who were not performing as they should. We decided after various meetings that Wendy should be given the position, and although there was some further minor development and training necessary, we felt that once she was in the job her behaviour would become even more 'leader-like'. In collective and ascriptive cultures you speak out when you have a position which says that you are allowed to. When Wendy had the legitimate title of departmental manager then she behaved with more power and authority, because the position allowed her to.

(4) *Connect.*   What face-to-face approach needs to be taken? The general connecting lessons here are about how to ensure that people are comfortable before asking them to make critical statements. David had said 'Does anyone have any criticisms of the way things are going in the department; how can I improve my leadership style?' When there was no reply, he repeated (in a louder tone of voice), 'Come on – it's OK – you know you can say what you

like to me, don't be shy.' Even more silence. To make his collective and ascriptive staff feel comfortable what he should have done was to divide them into five small groups, appointed a leader and asked them to come up with 10 ways to enhance departmental efficiency. He then should have left the room and come back in one hour. In the real situation, when David tried this he got 50 ideas about how to improve the department – and 15 of them were directly useable – and by implication the ideas were the feedback that he had been looking for about his leadership style. Isn't that better than the embarrassed silence which came from his directness, not to mention a continued succession of expat managers?

### Business Impact

One major impact of not recognizing the talent of the local leadership is the continued high cost of bringing out expat managers. Most contracts are for three years, and it is really only in the third year that expat managers are functioning at the highest and most efficient level, as they are then most comfortable with the culture and the job. Then they leave, and one has to go through the whole familiarization process again. The simple way to look at this is that with two expat managers, one only gets two out of six years in which they are 100 per cent fully functioning at the highest level.

Another major impact is on the loyalty of local staff. Asians tell me that when assessing a new job they look around to see how high is the highest Asian in the company. Many MNCs still have white faces above a certain level. Asians will still work for those companies but only for a certain amount of time until they have learnt all they can. Then they will take their skills and join a company where there is no limit to how high they can move in their career. In today's world where the importance of intellectual capital is more and more recognized, the cost of losing such local talent because of the lack of cultural literacy is enormous.

### Case 3: Joint Venture in Thailand

'Quest' is a major US multinational in the household electrical business and operates all over the world in joint ventures with local manufacturing partners. In most countries there is a parent company representative office which works with the local partner. Relations in Thailand between the local office and the joint venture partner have been strained of late due to poor human communications and misunderstanding. The representative office thinks that the partner has been slow and unduly sensitive to criticism, and the Thai partner is of the opinion that the representative office has been too impatient over unavoidable delays in marketing and distribution. Market share has slipped to their rival, also a major US multinational.

Paul is the new US country manager in Bangkok, and has been in the country for two months. He is on the 'fast track' and is on a 2-year assignment with the expectation that he turns things around in Thailand. Anxious to prove that he can 'get things on an even keel' he has arranged a workshop between the senior group from the local representative office (a mixture of Thais and US expatriates) and local partner (all Thais). The president of the local partner company, who is one of the most respected businessmen in Thailand, will also attend the workshop.

A meeting has been called to discuss the workshop. Present are:

- Sichai, aged 48, the Thai training manager from the joint venture: 25 years' service
- Greg, 32, Australian regional training manager for the parent company: based in Singapore, 5 years' service, 1 year in Asia
- Paul, 36, country manager of parent company
- Elizabeth, 35, the US facilitator. It is normal practice to bring out US facilitators, and Paul had insisted on Elizabeth, a consultant who has been a facilitator at the company's vision workshop in the US and been highly praised for her ability to get people to open up and discuss real problems. This is her second visit to Asia, having previously conducted a HK Workshop.

Under discussion is the simple matter of dress code. Elizabeth says 'It's casual dress code – no ties. Is that OK with you Sichai?' Sichai looked uncomfortable and after a while said, 'I'm not sure.' Greg jumped in, 'No, no – we need people to open up. Part of the problem in the JV has been the formality of the meetings. If we let people know that the dress code is informal, we're setting the right tone at the beginning. I agree no ties.'

A debate ensued about the importance of getting everyone to relax, with Sichai saying little.

*Applying the 4 Skills of Cultural Literacy*

(1) *Detect.*   What are the cultural problems? Clearly again communication style is a major issue here. Sichai is silent, and silence from an intelligent man with long experience in training and development is an indication that there is a cultural problem. Paul is new to the situation and so has not built up a relationship with Sichai so that Sichai feels uncomfortable to share his view.

(2) *Audit.* Which of the 7 cultural dimensions are influencing this situation?


- Universal–Particular
- Individual–Collective
- Affective–Neutral
- Achievement–Ascription


It should be obvious how Individual–Collective and Affective–Neutral are influencing the situation, so there is no need to discuss those at length. Universalism is showing itself loud and clear because all the players apart from Sichai bring their own behaviour from their own culture and assume that because 'casual' to them means jeans and t-shirt this is what means all over the world. The main dimensions influencing this situation are Achievement–Ascription, and they are influencing in three ways. Firstly as the president of the Thai partner is going to be there, then the Thai staff will feel uncomfortable if they do not dress formally, so Sichai knows that he cannot agree to a Western definition of casual. The second way is in the meeting itself. Sichai has his boss and his matrix boss in the meeting. If Sichai is to give his point of view he will have to disagree in public with his bosses and because of his cultural norms around seniority he is not likely to do this. The third way is more subtle. Sichai is the oldest person at the meeting and has responsibility for training, hence he should be treated with deference and respect. Paul invited Elizabeth without consulting with Sichai and he is not being accorded the status as the local training manager he deserves in this whole process.

(3) *Integrate.* The integration is done by paying attention to the 'ascriptive' culture's need for formality and making sure they can wear ties, and letting them know what they are expected to wear at other sessions. For the achievement cultures, informality is achieved by letting them know they can wear casual clothes in the evening sessions.

(4) *Connect.* What face-to-face approach needs to be taken? The face-to-face issues in this situation are very important. Greg should keep his opinion to himself and ask Sichai first what he thinks should be done. Once Greg expressed so forcefully that there should be no ties, Sichai could not give his real view as it would have created disharmony in the group. In the situation that developed the more Sichai is asked what his views are the more pressure he will feel. So the best thing to do is to close the meeting and let people know that at the first session ties should be worn out of respect for the Thai president.

*Business Impact*

One might say that this is a very small issue. What does it really matter what people wear? Often it is the small issues which are masking a very important cultural clash. The two sides are not seeing eye to eye in the production of Quest's products. The Americans do not understand the delays and the Thais feel resentful about the Americans' impatience. What you have in the example of the 'tie' is a symbolic reproduction of the problem with the whole partnership. The Westerners are insisting on their way of doing things and the Thais are being silent and resentful. What the Thais may feel is, 'If there are no ties then there will be no Thais' – i.e., the Thai way of doing things will not be respected. The tie issue is a perfect opportunity to sort out the issue in the larger partnership, and if it is not sorted out the workshop will repeat the problems and will continue with all the attendant costs of missed targets. When there are seemingly small issues, always take a closer look, for what you will see may give a valuable clue to sorting out much bigger issues. (For further discussion of Thai culture, see Holmes and Tangtongtavy, 1995.)

## CONCLUSION: CULTURAL LITERACY, ITS LINK TO BUSINESS SUCCESS IN ASIA-PACIFIC

I have tried to show the importance of cultural literacy to business success in Asia-Pacific. In the situations I work with, the executives are smart, intelligent people who have a commonsense understanding of cultural differences, but somehow do not feel comfortable about applying that knowledge in a vocal way to the problems they face. Mostly they say they that did not want to upset people by generalizing about the other culture, or sharing opinions which they feel are too subjective. Also there is a difficulty caused by the misperception of Western notions of diversity. Westerners say they are often uncomfortable about expressing their true feelings about cultural issues because 'diversity' programmes encourage 'political correctness'. Whilst applauding the benefits of many diversity programmes it is problematic to export the approach to cultures who see diversity in different ways. Political Correctness can often stifle debate about the very issues that they are set up to deal with. Diversity itself needs to be diverse. (For further discussion of this topic, see Roosvelt, 1996.) For these reasons a common language with which everyone feels comfortable is vital when companies want to solve cultural issues.

I would like to conclude by sharing some qualities which I think indicate success in multi-cultural situations in Asia-Pacific. Whilst I think that most of us have abilities in working with other cultures there are qualities and skills that I have noticed over the years which if present are likely to predict success when selecting staff to work in situations where cultural literacy is important.

**The Most Successful Culturally Literate Executives:**

- Have a 'Big Picture' view of organization rather than a focus on individual fragments
- Adapt well to new people, places, situations
- Strive for global standards and practices while allowing for local variations
- Do not want to change everything in the first six months
- Understand their own cultural values, and realize that their culture's way of doing things is not necessarily the best or the only way
- Communicate and collaborate well face to face
- Are skilled at reading what a person is not saying – body language
- Have a basic liking for local people and culture
- Make an effort to get to know local people and customs – are adventurous
- Listen well
- Have enjoyed living overseas
- Do not constantly 'whine' about the locals
- Have foreign-language skills
- Do not spend all the time with their own expatriate community
- Have good interpersonal skills in their own culture
- Have mastered the 4 Skills of Cultural Literacy!

**References**

Hampden-Turner, C. and Trompenaars, F. (1997) *Mastering the Infinite Game* (Oxford: Capstone).
Holmes, H. and Tangtongtavy, S. (1995) *Working with the Thais* (White Lotus Co. Ltd).
Leung, K. and Tjosvold, D. (1998) *Conflict Management in Asia Pacific* (Chichester: Wiley).
Roosvelt Jr., R. (1996) *Redefining Diversity* (AMACOM, a division of the American Management Association).
Shein, E. (1996) *Strategic Pragmatism* (Boston, MA: MIT Press).
Spekman, R. E., Isabella, L. A. and MacAvoy, T. C. (2000) *Alliance Competence. Maximizing the Value of Your Partnerships* (New York: John Wiley).
Trompenaars, F. (1993) *Riding the Waves of Culture: Understanding Cultural Diversity in Business* (London: The Economist Books).

# Index